Enantiopure Iminophosphonamide Complexes:
Synthesis, Photoluminescence and Catalysis

Enantiopure Iminophosphonamide Complexes: Synthesis, Photoluminescence and Catalysis

Zur Erlangung des akademischen Grades eines

DOKTORS DER NATURWISSENSCHAFTEN

(Dr. rer. nat.)

der KIT-Fakultät für Chemie und Biowissenschaften

des Karlsruher Instituts für Technologie (KIT)

vorgelegte

DISSERTATION

von

M.Sc. Bhupendra Goswami

aus

Rajasthan, India

KIT-Dekan: Prof. Dr. Manfred Wilhelm

Referent: Prof. Dr. Peter W. Roesky

Korreferent: Prof. Dr. Annie K. Powell

Tag der mündlichen Prüfung: 22.10.2020

Bibliografische Information der Deutschen Nationalbibliothek

Die Deutsche Nationalbibliothek verzeichnet diese Publikation in der Deutschen Nationalbibliografie; detaillierte bibliographische Daten sind im Internet über http://dnb.d-nb.de abrufbar.

1. Aufl. - Göttingen: Cuvillier, 2020

Zugl.: (KIT) Karlsruhe, Univ., Diss., 2020

The presented work was carried out in the period from 01.08.2017 to 09.09.2020 at Institute of Inorganic Chemistry in Karlsruhe Institute of Technology (KIT) under the supervision of Prof. Dr. Peter W. Roesky. This work was funded by KIT.

Nonnenstieg 8, 37075 Göttingen

Telefon: 0551-54724-0

Telefax: 0551-54724-21

www.cuvillier.de

1. Auflage, 2020

Gedruckt auf umweltfreundlichem, säurefreiem Papier aus nachhaltiger Forstwirtschaft.

ISBN 978-3-7369-7327-5

eISBN 978-3-7369-6327-6

Table of Contents

1 Introduction

1.1 Amidinates

Amidinates are a well-established class of monoanionic *N*-chelating ligand systems. The coordination chemistry of amidinate ligands has been reported with almost every metal or metalloid of the periodic table.[1] Amidinate anions with the general formula $[RC(NR^1)(NR^2)]^-$ can be considered as the nitrogen analogues of carboxylates $[RCO_2]^-$ (Figure 1.1). The ligand framework is rather versatile, as the substituents (R, R^1, R^2) present at the central carbon and nitrogen atoms can be widely varied, which allows a fine tuning of electronic and steric properties. Studies have revealed that, in order to stabilize compounds in elusive low-valent oxidation states, it was crucial to introduce a significant steric demand on the amidinate ligand backbone.[2] These interesting features of the amidinates make them contender of the ubiquitous cyclopentadienyl ligands.

Depending on the type of metals and substituents present on the central carbon and nitrogen atoms of amidinates, a variety of coordination modes with main group elements, transition metals, lanthanides and actinides are observed (B, Figure 1.1).[1a,3] Bulky substituents on the central carbon and nitrogen atoms typically induce a monodentate (κ^1) coordination mode (a),[4] although this mode is relatively rare in general.[5] The chelating (κ^2) coordination mode, in which delocalization of the double bond is commonly observed (b), forms a four-membered cycle with the coordinated element.[6]

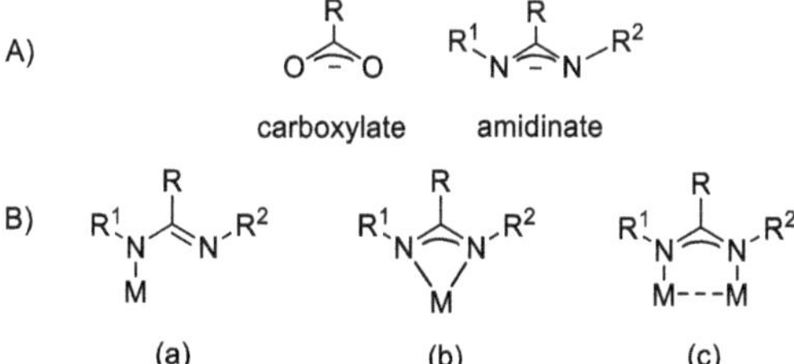

Figure 1.1 (A) Carboxylate and amidinate ligand frameworks, (B) possible coordination modes of amidinates.

The bridging coordination mode (c) is often found in paddlewheel-type coordination compounds with the general formula of $M_2(\text{amidinate})_n$ (n = 2-4), in which three or four amidinates bridge two metal centres.[7] The potential for metal-metal interaction and the fascinating chemistry of these paddlewheel coordination compounds have been extensively investigated and are still an actively pursued research field.[8] Owing to very small coordination pockets of amidinates, the latter offer very small bite angles (N-M-N angles) which lie in the range of 63-65°.[9]

Early reports have shown that amidinate metal complexes can be synthesized by heating metal carbonyls with a corresponding amidine.[10] Currently, the most widespread and routinely used methods include salt metathesis reactions between metal halides and alkali metal amidinates.[3] Deprotonation of amidines with metal alkyls or amides has also been used as an alternative approach to salt metatheses.[4b,11] However, this method is not commonly used which can be traced back to the low availability and stability of many metal alkyls. Additionally, direct insertion of carbodiimides (R-N=C=N-R') into M-C bonds has also been employed to synthesize metal amidinate complexes.[12]

The versatility of amidinates in coordination chemistry has led to a plethora of metal complexes and many of them have been applied in homogeneous catalysis or in material science to design novel materials.[12] Examples of application can be found in chemical vapor deposition (CVD), which requires metal precursors without any carbon in the direct coordination sphere in order to minimise carbide formation. Therefore, only coordinated by nitrogen atoms, the amidinate complexes of divalent Mg (Figure 1.2 (a)), trivalent Al and Ga have been used as CVD precursors to prepare metal nitride semiconductors (Figure 1.2 (b)).[13] As another example, cationic group 13 metal amidinate complexes have been utilized for olefin polymerization (Figure 1.2 (c)).[14] Prominent for the stabilization of reactive low-valent group 14 compounds, the amidinate ligand $[PhC(^tBuN)_2]^-$ was successfully used to access stable Si(II), Ge(II) and Sn(II) species (Figure 1.2 (d)).[15] More recently, our group has reported the first silicon analogue of the aromatic cyclobutadiene dication, $[\{PhC(^tBuN)_2Si\}_4](BPh_4)_2$ in which the four silicon centres adopt an almost square planar geometry.[16] Transition metal complexes of certain amidinates have been utilized as precursors in atomic layer deposition (ALD) where the crucial point is the volatility of the metal complexes, which can be controlled by taking advantage of substituent variation in the amidinate framework (Figure 1.2 (e)).[17] Moreover, the coordination chemistry of amidinates with rare earth metals was pioneered by the group of Edelmann more than two decades ago.[18] The amidinate ligand have been used to stabilize all available oxidation states (+2, +3, +4) of different lanthanides.[12]

In this regard, a series of homoleptic lanthanide complexes with various lanthanide metals have been reported in early studies using the $[RC_6H_4C(NSiMe_3)_2]^-$ amidinate anion (Figure 1.2 (f)).

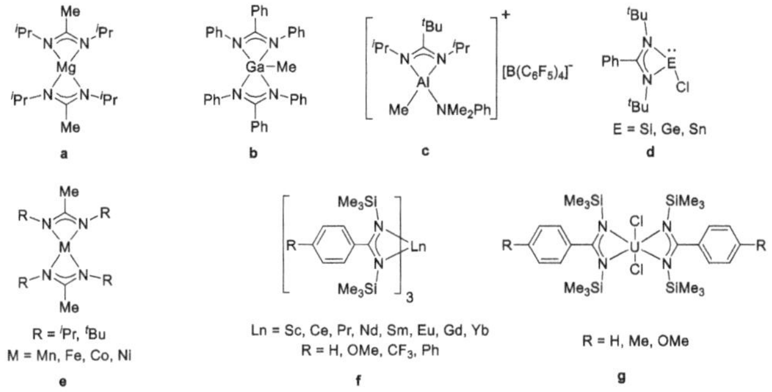

Figure 1.2 Selected examples of the amidinate-coordinated compounds.[12]

Since then, several other homoleptic lanthanide complexes have been synthesized and utilized as catalysts in the ring opening polymerization (ROP) of ε-caprolactone.[19] The most exciting application of a homoleptic lanthanide complex is the use of the erbium tris-amidinate complex $[\{MeC(N^tBu)_2\}_3Er]$ as a dopant source to prepare silicon nanocrystals.[12] The formation of mono-, bis-, or tris-amidinate substituted lanthanide complexes depends on the steric demand of the respective ligand and the ionic radius of the lanthanide, as described by the group of Teuben.[20] Additionally, the versatility of amidinate ligands can also be seen in the stabilization of actinide(IV) complexes and their use in ring opening polymerization of ε-caprolactone (Figure 1.2 (g)).[21]

1.2 Iminophosphonamide

In the monoanionic amidinate ligands described above (Section 1.1), a carbon corresponds to the central atom of the N-X-N motif. This central atom can be replaced by other elements and thus, a variety of N-X-N type bidentate chelating ligands are possible. Among those N-X-N ligands, boraamidinate (A), triazenide (B), sulfinamidinate (C) and iminophosphonamide (D) systems are known in the literature, where X = BR,[22] N,[23] SR,[24] and PR_2,[25], respectively (Figure 1.3). Not surprisingly, altering the central atom X, the electronic properties of the resulting ligands vary significantly.

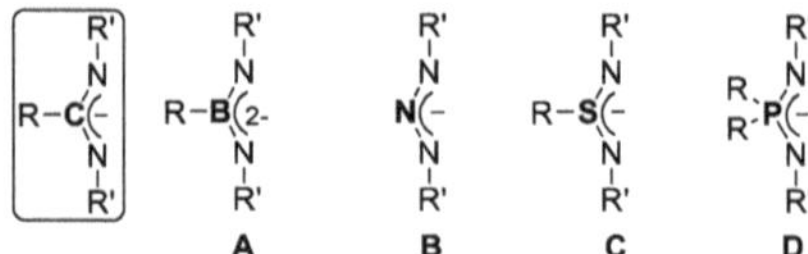

Figure 1.3 Various analogues of the amidinate ligand system.

At first glance, the coordination chemistry of these derived ligand systems can be expected to be similar. This is however not the case, as the respective bond lengths and bite angles differ, which has a considerable impact on the chemical behaviour in coordination compounds. Among the different classes of heteroatomic N-X-N type ligands, iminophosphonamides (X = PR_2) offer the longest X-N bond (1.60 Å), which leads to a wider bite angle in corresponding metal complexes. Furthermore, the phosphorus atom in the backbone of the iminophosphonamide ligand system serves as a spectroscopic "marker" for ^{31}P NMR spectroscopy, which allows a convenient monitoring of the reactions when using these ligands. In addition to this, the steric demand on the central X atom of the iminophosphonamide ligand backbone is comparatively greater than that in other NXN frameworks.

The iminophosphonamide ligand system $[(R_2P(NR')_2]^-$ can be considered as an analogue of the phosphinate anion ($R_2PO_2^-$) in which the oxygen atoms have been replaced by two amide groups. Three resonance forms can be depicted for this system (Figure 1.4). The nitrogen atoms in iminophosphonamides carry a negative charge and are σ-bonded to the positively charged phosphorus centre, thus making the NPN system a relatively stable zwitterionic framework.[26]

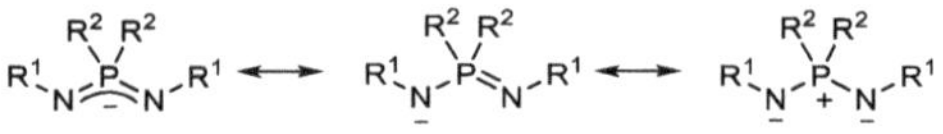

Figure 1.4 Resonance forms of the iminophosphonamide.

In general, iminophosphonamides can be synthesized by following two methods (Scheme 1.1):

(1) Kirsanov condensation: a reaction is carried out between tertiary phosphines R_2PX (X = Cl, Br) and primary amines ($R'NH_2$) leading to the formation of a diamino phosphonium salt, which, by using a strong base, is converted to the corresponding iminophosphonamide.[27]

(2) Staudinger reaction: a phosphonamine (R_2PNHR') or phosphine (R_2PH) is reacted with a stoichiometric amount of an organic azide ($R'N_3$), which directly leads to the protonated iminophosphonamine ligand *via* release of N_2.[28]

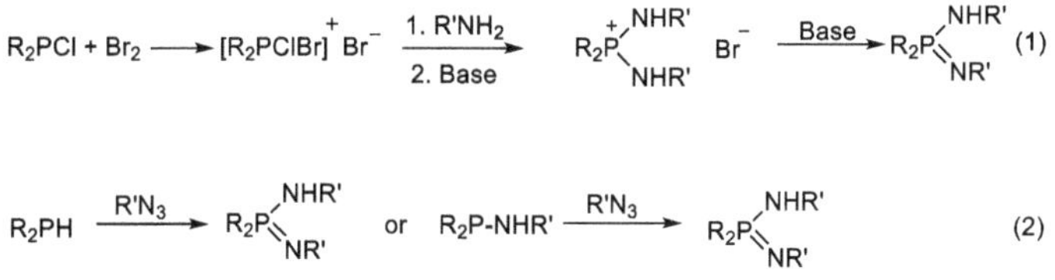

$$R_2PCl + Br_2 \longrightarrow [R_2PClBr]^+ Br^- \xrightarrow[\text{2. Base}]{\text{1. R'NH}_2} [R_2P(NHR')_2]^+ Br^- \xrightarrow{\text{Base}} R_2P(NHR')(=NR') \quad (1)$$

$$R_2PH \xrightarrow{R'N_3} R_2P(NHR')(=NR') \quad \text{or} \quad R_2P\text{-}NHR' \xrightarrow{R'N_3} R_2P(NHR')(=NR') \quad (2)$$

Scheme 1.1 Iminophosphonamide synthesis via Kirsanov (1) and Staudinger reaction (2).

Based on the above-mentioned methods, a variety of iminophosphonamines and their metal complexes has been reported in the literature. The respective metal complexes can be obtained either by salt metathesis reactions between alkali metal salts and metal halides or by deprotonation reactions (Scheme 1.2). Also, similarly to amidinates, iminophosphonamide ligands can also show the three types of coordination modes depicted in Figure 1.1.

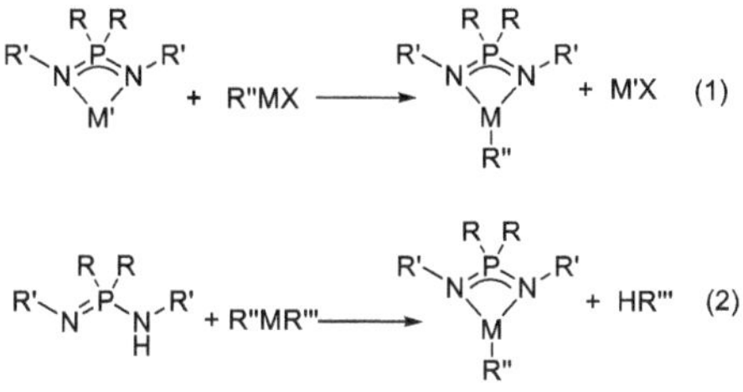

$$[R_2P(NR')_2]M' + R''MX \longrightarrow [R_2P(NR')_2]MR'' + M'X \quad (1)$$

$$R'N{=}PR_2{-}N(H)R' + R''MR''' \longrightarrow [R_2P(NR')_2]MR'' + HR''' \quad (2)$$

Scheme 1.2 Typical reaction pathways to synthesize iminophosphonamide metal complexes.

1.2.1 s-Block metal complexes of iminophosphonamides

Alkali metal iminophosphonamide complexes are useful precursors for salt metathesis reactions. Treatment of the protonated iminophosphonamines with alkali metal bases such as *n*BuLi, NaH, $[Na\{N(SiMe_3)_2\}]$ or KH, $[K\{N(SiMe_3)_2\}]$ leads to the corresponding alkali metal complexes. A variety of alkali metal complexes have been reported and structurally characterized using various iminophosphonamines such as $[(2,6\text{-}i\text{Pr}_2C_6H_3N)P(Ph_2)(Nt\text{Bu})]H$,[25c] $[\{(2,6\text{-}i\text{Pr}_2C_6H_3N)_2P(Ph_2)\}]H$, $[\{(2,4,6\text{-}Me_3C_6H_2N)_2P(Ph_2)\}H]$,[26,29] $[(2,4,6\text{-}Me_3C_6H_2N)_2P(Ph_2)]H$ and $[\{(2,6\text{-}i\text{Pr}_2C_6H_3NH)P(Ph_2){=}N(C_6H_2\text{-}2,4,6\text{-}Me_3)\}]$,[30] *rac*-[*trans*-1,2-$C_6H_{12}\{NP(Ph_2)N(Ar)\}_2$].[31] Based on the structural investigation of the reported complexes, it can be anticipated that the aggregation state of these metal complexes depends on the N-substituents and the ionic radii of the central metal atoms. For example, the potassium salt [K{(2,6-*i*$Pr_2C_6H_3N)_2P(Ph_2)$}]$_n$ is polymeric in the solid state (Figure 1.5). Surprisingly, the potassium ion only shows coordination with the arene rings from two Dipp (Dipp = 2,6-*i*$Pr_2C_6H_3$) substituents and two phenyl groups from a neighbouring ligand in the polymeric chain, instead of realizing one of the typical bonding motifs shown in Figure 1.1.

Figure 1.5 The polymeric chain of [K{(2,6-*i*$Pr_2C_6H_3N)_2P(Ph_2)$}].[29]

In addition, the alkali metal complexes of the {$Ph_2P(Me_3SiN)_2$}$^-$ ligand show non-uniform coordination patterns throughout the series from Li to Cs.[32] These complexes were synthesized by deprotonation reactions of {$PPh_2(Me_3SiN)_2$}H with *n*-BuLi, NaH, KH, Rb, and Cs (Scheme 1.3). The solid-state structures reveal that the lithium and potassium salts [{$Ph_2P(NSiMe_3)_2$}Li(thf)$_2$] and [{$Ph_2P(NSiMe_3)_2$}K(thf)$_4$] exist in monomeric forms. The sodium analogue shows an unusual, highly reactive example of a low melting sodium/sodiate ion pair [{$Ph_2P(NSiMe_3)_2$}$_2$Na][Na(thf)$_6$]. However, the corresponding Rb and Cs metal complexes exist in dimeric [{$Ph_2P(Me_3SiN)_2Rb(thf)$}$_2$] and [{$Ph_2P(Me_3SiN)_2Cs$}$_2$] forms.

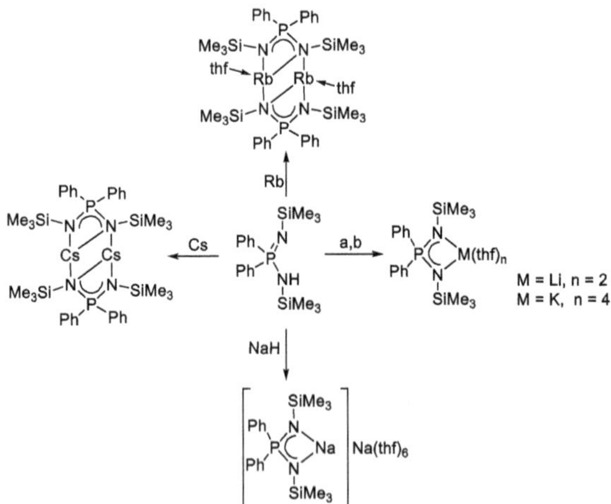

Scheme 1.3 Synthesis of alkali metal iminophosphonamide complexes; (a) *n*-BuLi, (b) KH.[32]

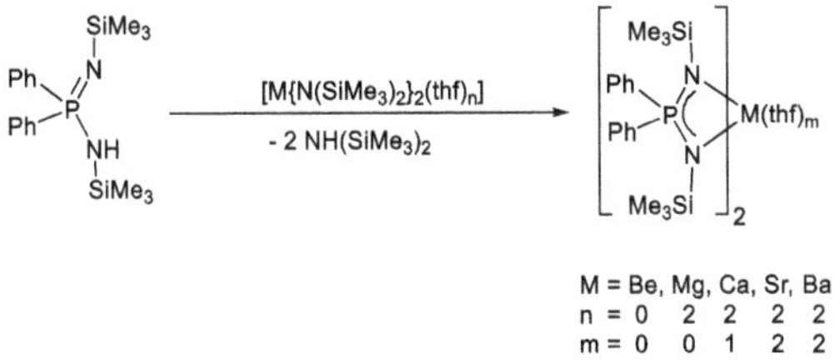

Scheme 1.4 Synthesis of alkaline earth metal iminophosphonamide complexes.[33]

Alkaline earth metal complexes supported by iminophosphonamides are typically synthesized by deprotonation of the corresponding iminophosphonamines by alkaline earth metal amides, $[M\{N(SiMe_3)_2\}_2]$. In contrast to the alkali metal complexes based on the $\{Ph_2P(Me_3SiN)_2\}^-$ ligand, the corresponding group-2 metal (M = Be, Mg, Ca, Sr, and Ba) complexes show a uniform coordination pattern (Scheme 1.4).[33] The common structural feature of all of these complexes is the coordination of two monoanionic ligands to the metal centre to form a mononuclear iminophosphonamide complex $[\{Ph_2P(NSiMe_3)_2\}_2M]$. Depending on the ionic radius of the respective metal, the incorporation of additional thf molecule(s) to satisfy the metal coordination sphere is observed. By utilizing sterically demanding iminophosphonamide ligands such as $\{(2,6\text{-}^iPr_2C_6H_3N)_2P(Ph_2)\}^-$, highly reactive $[Mg]^{+1}$ complexes have also been reported.[34] More recently, this complex was utilized to study the mechanistic insights of anionic ligand exchange and fullerene reduction.[35]

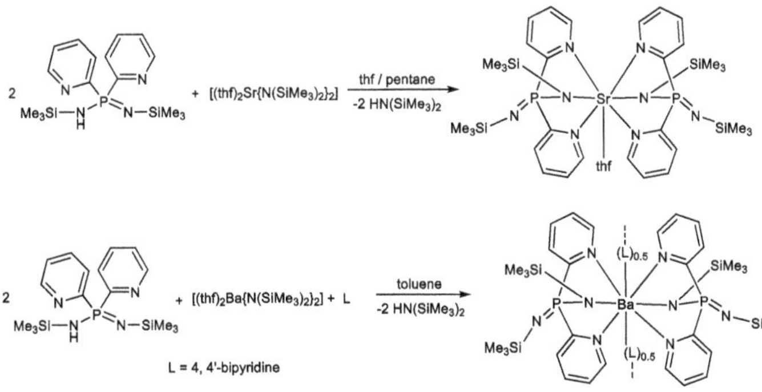

Scheme 1.5 Alkaline earth metal complexes based on *Janus head* iminophosphonamides.[36]

In order to introduce additional coordination sites in the $[Ph_2P(Me_3SiN)_2]^-$ iminophosphonamide ligand, a new *Janus head* $[Py_2P(Me_3SiN)_2]^-$ ligand has been reported where the phenyl groups on the phosphorus have been replaced by pyridyl substituents.[36]

This multi-dentate *Janus head* ligand exhibits two possible chelating coordination pockets based on (a) the NPN functionality and (b) two remote pyridyl substituents.

The reaction of the protonated Janus-type ligand with [M{N($SiMe_3$)$_2$}$_2$] (M = Sr or Ba) proceeds smoothly at room temperature, leading to the formation of the [{$Py_2P(NSiMe_3)_2$}$_2$Sr(thf)] and [{$Py_2P(NSiMe_3)_2$}$_2$Ba(4,4'-bipy)]$_n$ complexes, where the solid state structure of barium complex show a zigzag polymeric arrangement assisted by 4,4'-bipyridine ligands.(Scheme 1.5).[36]

1.2.2 p-Block: group 13 and 14 compounds of the iminophosphonamides

1.2.2.1 Iminophosphonamide complexes of group 13 metals

Complexes containing group 13 metals have attracted a long-standing interest in various homogeneous catalyses involving polar substrates and have promoted Lewis acid mediated activation reactions.[37] In this regard, metal complexes containing M-X bonds (X = halide, alkyl, and hydride) are widely used catalysts in various reactions such as cationic polymerization,[14] hydroboration,[38] hydro functionalization,[39] Friedel–Craft and Diels–Alder reaction,[40] and epoxidation of alkenes.[41]

Typically, iminophosphonamides bearing medium-sized substituents show a chelating bonding mode with the group 13 metals and form four-membered metallacyclic rings. For example, Roesky *et al.* reported on aluminium complexes of the iminophosphonamide {($Ph_2P(Me_3SiN)_2$}$^-$. These complexes were obtained *via* salt metathesis reactions between the respective *in situ* generated lithium iminophosphonamide with group 13 metal precursors or *via* a direct base elimination reaction involving the iminophosphonamine {($Ph_2P(Me_3SiN)_2$}H (Scheme 1.6).[42]

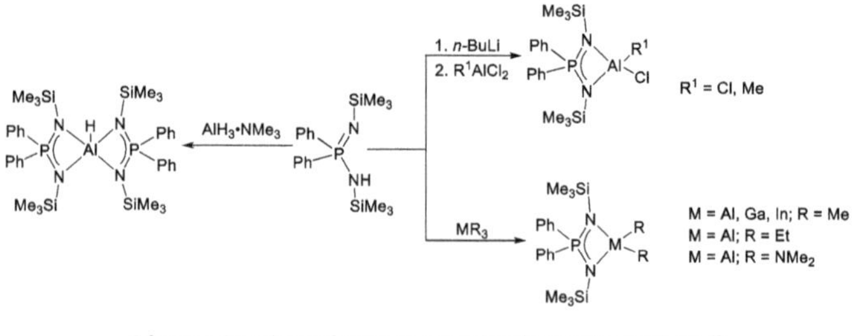

Me3Si SiMe3 Ph N H N Ph P Al P Ph Ph N N Me3Si SiMe3 ← AlH3•NMe3 — Ph2P(NSiMe3)(NHSiMe3) → 1. n-BuLi 2. R¹AlCl2 → Ph2P(NSiMe3)2Al(R¹)Cl; R¹ = Cl, Me → MR3 → Ph2P(NSiMe3)2MR2; M = Al, Ga, In; R = Me; M = Al; R = Et; M = Al; R = NMe2

Scheme 1.6 Synthesis of group 13 iminophosphonamide complexes.[42]

Among the synthesized compounds, the tetracoordinated dimethyl complexes $[\{Ph_2P(NSiMe_3)_2\}Me_2M]$ (M = Al, Ga, In) are thermally stable and show defined melting points. The monomeric nature of these metal complexes was confirmed by measuring the corresponding cryoscopic molecular weight.[42b] The aluminium centre in the aluminium monohydride complex is pentacoordinated and features a distorted trigonal bipyramidal geometry. In all of these complexes, the P-N contact lengths are in between single and double bonds.[43] Apart from the above mentioned examples, only few group 13 metal complexes based on other NPN backbones, e.g. $\{(2,6\text{-}^i Pr_2C_6H_3N)Ph_2P(N^tBu)\}^-$,[25c,44] $\{(2,6\text{-}^i Pr_2C_6H_3N)_2PH(^tBu)\}^-$,[45] and *rac*-[*trans*-1,2-$C_6H_{12}\{NP(Ph_2)N(Ar)\}_2]^-$ (where Ar = 2,4,6-$Me_3C_6H_2$ or 2,6-$Me_2C_6H_3$),[1b] have been structurally characterized. Interestingly, the sterically demanding iminophosphonamide $\{(2,6\text{-}^i Pr_2C_6H_3N)_2Ph_2P\}^-$ ligand has been used to stabilize heavy monovalent group 13 (Ga(I), In(I), Tl(I)) metal complexes.[46]

1.2.2.2 Iminophosphonamide compounds of group 14 elements

Heavy divalent group 14 compounds are collectively termed as tetrylenes [R_2E:] with (E = Si, Ge, Sn and Pb with R = supporting ligand). In 1974, the group of Lappert reported that these compounds can be successfully stabilized by sterically demanding amido ligands (-NR_2) in which the lone pair of the amido group stabilizes the vacant p-orbital of the divalent group 14 elements by an effective P_π-P_π interaction.[47] Following the first report of amidinate-stabilized divalent group 14 compounds $[\{(PhC(^tBuN)_2\}ECl]$, (E = Si, Ge, Sn), the corresponding chemistry has been expanded drastically over the last two decades.[15b,15c,48]

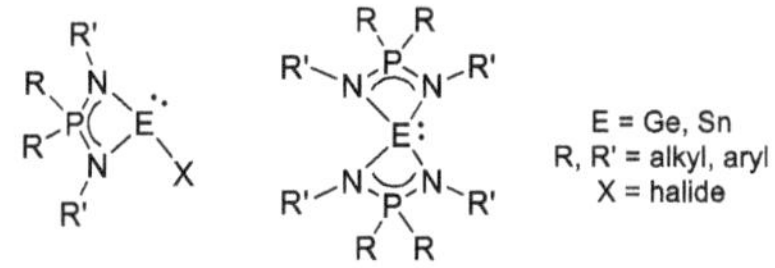

E = Ge, Sn
R, R' = alkyl, aryl
X = halide

Figure 1.6 Literature-known tetrylenes based on iminophosphonamides.[49]

Compared to the well-established chemistry of amidinate-stabilized tetrylenes, only a few examples of tetrylenes based on iminophosphonamide ligands have been reported (Figure 1.6). Such examples include the germylene $[Ph_2P(N\mathit{t}Bu)(2,6\text{-}\mathit{i}Pr_2C_6H_3N)GeX]$ (where X = Cl, O^{t}Bu, OTf)[49a] and stannylene $[\{Ph_2P(NSiMe_3)_2\}_2Sn]$ and $[\{(Ph)(R)P(N\mathit{t}Bu)_2\}SnX]$ (where R = CH_2Ph or 8-C_9H_6N; X = Cl or $N(SiMe_3)_2$) compounds, which have been structurally

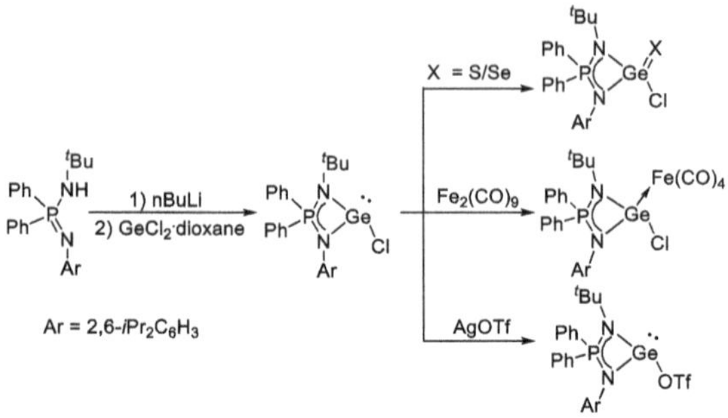

Scheme 1.7 Synthesis of an iminophosphonamide monochloride germylene and its reactivity.[49a]

characterized.[49-50] The general synthetic route for the preparation of the aforementioned low-valent compounds involve salt metathesis reactions using tetrylene halides. In this regard, the germylene monochloride [{(2,6-*i*$Pr_2C_6H_3N$)P(Ph_2)(N*t*Bu)}GeCl] can be synthesized by treatment of the *in situ* generated lithium salt of the ligand, **L**Li [**L** = (2,6-*i*$Pr_2C_6H_3N$)P(Ph_2)(N*t*Bu)], with one equivalent of $GeCl_2$·dioxane.[49a] The reactivity of the synthesized germylene monochloride was further studied towards oxidation, formation of adducts with transition metals and in salt metathesis reactions (Scheme 1.7). Additionally, the tetracoordinated iminophosphonamide complexes of Sn(II) and Pb(II), [{$Ph_2P(NSiMe_3)_2$}$_2$E] (where E = Sn, Pb), have also been reported.[49c] These homoleptic tetrylenes adopt a distorted trigonal bipyramidal geometry in which the respective lone pair occupies the equatorial position (Figure 1.6).

1.2.3 Transition metal complexes of the iminophosphonamide ligands

Transition metal complexes based on amidinate (NCN) ligands have been extensively studied, with reported catalytic applications in various organic transformations.[51] In contrast to the well-established chemistry of amidinate ligands, the corresponding chemistry of transition metal iminophosphonamides (NPN) is still in its infancy.[52] Nevertheless, in the past few decades, various transition metal complexes of iminophosphonamides have been employed as catalysts, e.g. in alkene polymerizations,[53] cyclopropanation reactions,[54] and in olefin oligomerizations.[55] In 1981, the group of Goddard reported the first nickel and palladium complexes starting with an (amino)bis(imino)phosphorane ligand (Scheme 1.8). The formation of these complexes occurred *via* unexpected allyl rearrangement of bis(η^3-allyl) Ni/Pd complexes involving the (amino)bis(imino)phosphorane ligand, leading to the formation of a σ-bonded P-allyl group. These complexes serve as an active catalyst for the polymerization of ethylene at 70 °C and 50 bars. Furthermore, various Ni(II) complexes, such as [{iPr$_2$P(NMe)$_2$}$_2$Ni], [{Ph$_2$P(NC$_6$H$_2$-2,4,6-Me$_3$)$_2$}$_2$Ni], and [{Ph$_2$P(NPh)$_2$}$_2$Ni] have also been structurally characterized.[30,56]

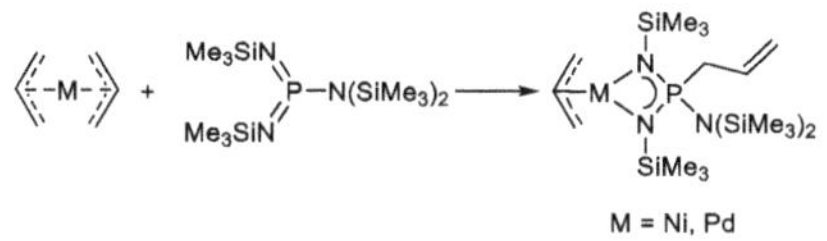

Scheme 1.8 Synthesis of nickel and palladium iminophosphonamide complexes.[53a]

In 1999, the group of Hofmann reported a copper(I) olefin complex stabilized by a sterically demanding [^{t}Bu$_2$P(NSiMe$_3$)$_2$]$^-$ ligand (Scheme 1.9).[57] The stability of this Cu(I) olefin complex is attributed to the small bite angle (N-Cu-N 77.80(9)°) which eventually enhances the π-back bonding towards the olefin and, thus, creates a strong dative bond between Cu(I) and the olefin. Utilizing this copper(I) olefin complex, the group of Hofmann was able to identify a Cu carbene complex relevant to catalytic cyclopropanation.[54b] Furthermore, the reactivity of this Cu(I) olefin complex has also been reported towards O_2 gas, leading to the formation of a neutral butterfly bis(μ-oxo) copper(III) species. Additionally, a salt metathesis reaction between CuCl and [{Ph$_2$P(NSiMe$_3$)$_2$}Li] afforded the only reported binuclear copper complex based on an iminophosphonamide ligand.[58] The corresponding crystal structure shows that the two Cu(I) centres are linearly coordinated *via* a bridging coordination mode of the two iminophosphonamides. The Cu-Cu distance is 2.62 Å and thus in the range of cuprophilic interactions.[58-59]

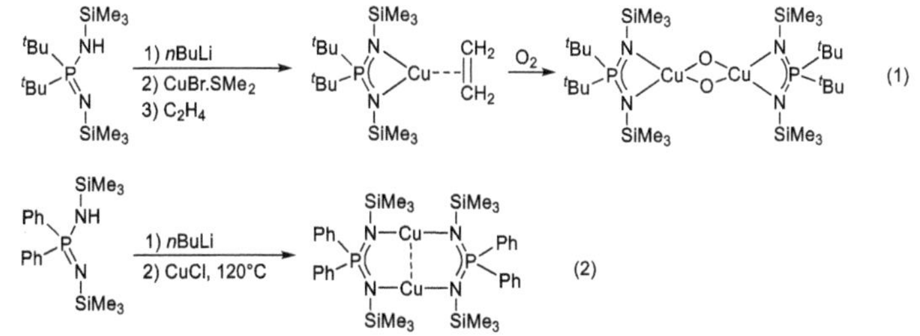

Scheme 1.9 Iminophosphonamide-supported copper(I) olefin and bis(μ-oxo) copper(III) species (1) and binuclear copper complex (2).[57]

Only one structurally-characterized example of a Cu(II) iminophosphonamide complex has been reported and was accessed by deprotonation of [{(2,4,6-Me$_3$C$_6$H$_2$N)$_2$P(Ph$_2$)}]H with

$[Cu\{N(SiMe_3)_2\}_2]$, resulting in the formation of the mononuclear homoleptic complex $[\{(2,4,6\text{-}Me_3C_6H_2N)_2P(Ph_2)\}_2Cu]$.[30] Corresponding divalent zinc complexes have only been obtained using the $\{(2,6\text{-}i Pr_2C_6H_3N)_2P(Ph_2)\}^-$,[26] and $\{Py_2P(Me_3SiN)_2\}^-$, NPN skeletons.[60] In this regard, the group of Stasch has demonstrated that a monovalent zinc complex could be stabilized by the sterically demanding $\{(2,6\text{-}i Pr_2C_6H_3N)_2P(Ph_2)\}^-$ ligand. Apart from these examples, Ru,[61] Co,[62] Pt,[53a,63] and Ag,[64] complexes based on iminophosphonamides have also been reported and structurally characterized.

In addition, the coordination chemistry of NPN ligands has also been investigated within early transition metals.[55,65] In 2011, Albahily *et al.* reported a remarkable chromium complex based on the sterically demanding ligand $\{Ph_2PN(^tBu)_2\}^-$.[66] This complex served as an active catalyst for selective trimerization of ethylene. Furthermore, group 4 metal complexes based on mono or di substituted iminophosphonamide ligands, such as $[\{Ph_2P(NSiMe_3)_2\}MCl_3(CH_3CN)]$ (M = Ti, Zr),[65a,65c] and $[\{R_2P(NR')_2\}_2MCl_2]$ (M = Ti, Zr),[65a,67] respectively, and the mixed metallocenes $[\{R_2P(NR')_2\}(Cp)MCl_2]$ (M = Ti, Zr, Hf)[67c] involving one Cp ring and one iminophosphonamide ligand have been reported. Out of these examples, the disubstituted and mixed metallocene complexes were found to be active catalysts for the polymerization of ethylene.

1.2.4 Rare earth metal complexes based on iminophosphonamides

Considering the rather well-studied amidinate chemistry of rare earth metals, the corresponding coordination chemistry with iminophosphonamides is relatively less explored. Pioneers in this field are Edelmann and Schumann, who contributed strongly to this research area since the 1980s.[68] Various NPN systems substituted at the *N*-centres by silyl, aryl and alkyl groups have been applied as ligands in rare earth metal chemistry so far (Figure 1.7).

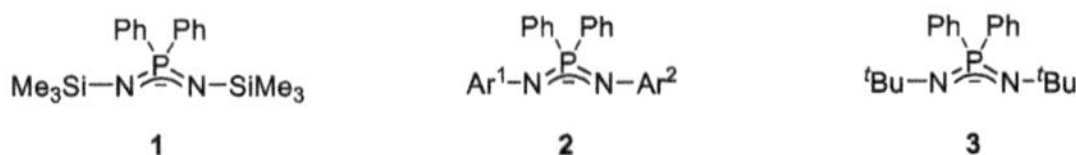

Figure 1.7 Iminophosphonamide ligands used in rare earth metal complexes.

Utilizing *N*-silyl substituted ligands, a series of Ln-COT (COT = cyclooctatetraene) NPN-complexes (Ln = Ce, Pr, Nd and Sm) have been reported and structurally characterized (**A**, Figure 1.8).[69] In 1991, the group of Edelmann synthesized Sm(III) and Yb(II) complexes bearing the same NPN type ligand (**B** and **C**, Figure 1.8).[70] The alkyl and aryl substituted lanthanide complexes of iminophosphonamides are mostly investigated as catalysts in the polymerization of olefins.[31,71]

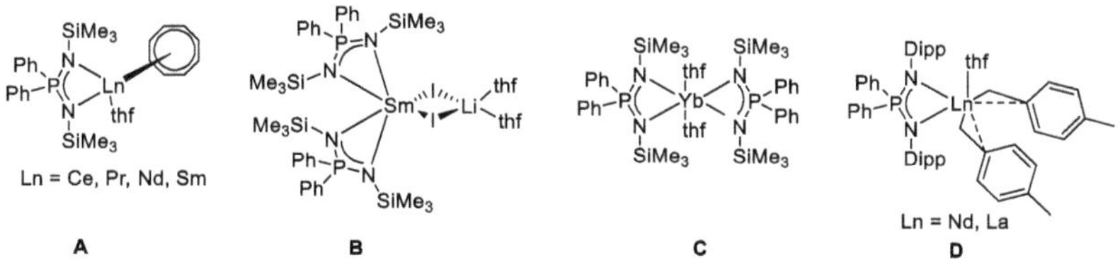

Figure 1.8 Selected examples of rare earth metal complexes based on NPN ligands (A-D).

Studies by the group of Cui *et al.* on the catalytic polymerization of isoprene showed that the catalytic activity and selectivity of rare earth metal complexes with functionalized NPN ligands depend on the substituents located on the nitrogen atoms, and the atomic radii of the respective central metal atoms. Sterically demanding ligands tend to favour a high 3,4-regio and stereoselectivity for the polymerization of isoprene. Furthermore, 4-methylbenzyl neodymium and lanthanum complexes (**D**, Figure 1.8) showed efficient catalytic activity in the *trans*-1,4-selective polymerization of isoprene.[72]

Alkyl or aryl substituted iminophosphonamide complexes were synthesized by reacting rare earth metal tris-alkyls with the corresponding iminophosphonamine, as reported by Hessen *et al.*[73] Among the above mentioned three types of NPN backbones, the rare earth metal chemistry of *N*-alkyl ligands (Figure 1.7, type 3) was first investigated by the group of Sundermeyer in 2016.[74]

For example, the synthesis of $\{Ph_2P(N^tBu)_2\}^-$ substituted dialkyl rare-earth metal (Sc, Y) complexes are shown in Scheme 1.10.

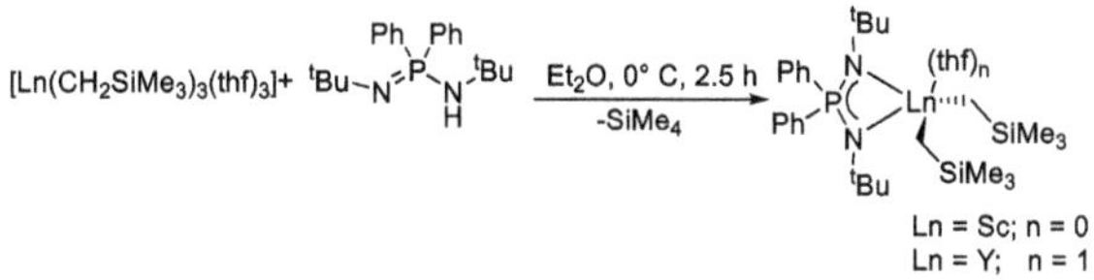

Scheme 1.10 Synthesis of dialkyl lanthanide complexes.[74]

Furthermore, the synthesis of mono-alkyl rare earth metal complexes could be achieved by using the appropriate stoichiometric ratio of the precursors. The mono-alkyl substituted complexes are reactive towards abstraction of acidic -CH protons (Scheme 1.11).[74]

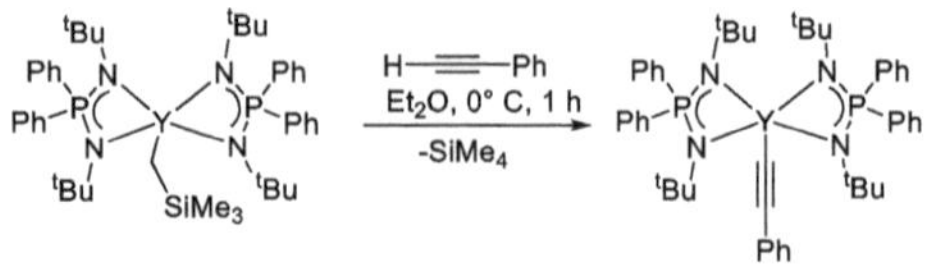

Scheme 1.11 Synthesis of alkynyl lanthanide complexes.

1.3 Enantiopure analogues of the amidinate and iminophosphonamide ligands

In the field of pharmaceutical industries, enantiopure or chiral products play a very important role in the synthesis and development of drugs. Based on the interaction of chiral drugs with biological targets such as proteins and nucleic acids, one enantiomer may act as a medicine whereas other one may show toxic effects.[75] Therefore, enantiopurity of a molecule is important for effective drug design. Enantioselective catalysis is an important tool to design the desired molecule. The progress in this field requires the design of new chiral ligands, which coordinate to the central metal ions and govern the enantioselectivity in asymmetric reactions.

In contrast to the above described, well-explored chemistry dealing with achiral *N*-chelating ligands, the corresponding chemistry involving chiral counterparts is much less developed. Prior to pioneering work of Roesky *et al.* concerning chiral amidinates in rare earth metal chemistry,[76] most of the published examples featuring chiral amidinates are shown in Figure 1.9 (**A**-**C**). In this regard, type **A** chiral amidinate ligands and the corresponding group 4 metal complexes were reported by the group of Eisen and revealed catalytic activity in polymerization of propylene.[77] In 2006, Togni *et al.* reported on a novel chiral ferrocene-based (type **B**) ligand and further presented a Rh complex coordinated by this ligand.[78] Group 4 metal complexes based on type **C** chiral amidinate have been reported by Li and co-workers.[79] In 1980, Brunner *et al.* have reported a type **D** (R = Ph) chiral amidinate, corresponding to the *N,N'*-bis-(1-phenylethyl)benzamidine framework and referred to as (*S*)- and (*R*)-HPEBA.[80] In 2006, Yashima *et al.* utilized the carbodiimide route to synthesize HPEBA.[81] However, this synthetic route requires laborious purification of the products by column chromatography. Our group reported an improved synthesis for the chiral HPEBA and its lithium and potassium salts (LiPEBA and KPEBA, respectively) in 2011.[48b] Further utilizing this chiral ligand, our group reported on the first chiral rare earth amidinate complexes and their catalytic activity and enantioselectivity in hydroamination reactions.[76]

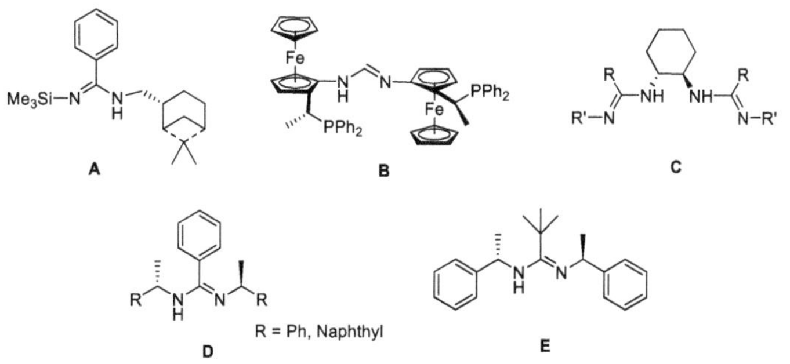

Figure 1.9 Examples of chiral amidine ligands.

Based on the nature of the metal centre and the substituents present on the ligand backbone (Figure 1.9 (**D**, **E**)), a variety of luminescent and catalytically active complexes, especially in hydroamination, were obtained.[82] Furthermore, the effect of chiral ligand in the magnetic behaviour of the lanthanide complexes has also been studied.[83]

It is known that some ligands induce more or less selectivity. Therefore, screening of other ligand is necessary in order to understand asymmetric induction. In this connection chemistry dealing with chiral iminophosphonamides is even less explored as compared to that of chiral amidinates. Only a few corresponding chiral ligands are known in the literature (Figure 1.10). The common feature of all these chiral iminophosphonamines is that only one nitrogen atom of the NPN backbone possesses a stereocenter. In this regard, Li *et al.* reported differently functionalized chiral mono(iminophosphonamines) (**F** and **G**, Figure 1.10).[84]

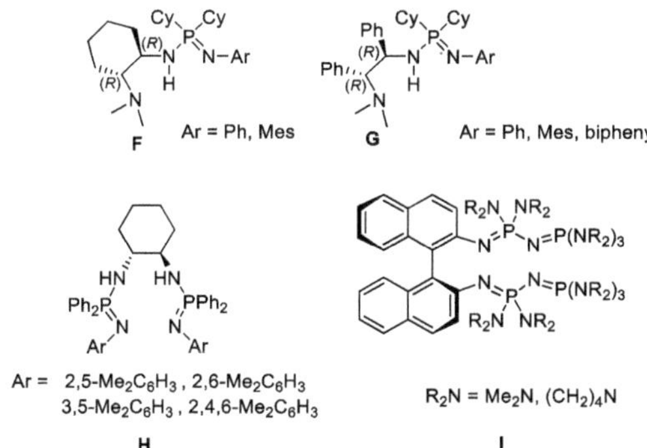

Figure 1.10 Types of chiral iminophosphonamine ligands known.

The corresponding yttrium and alkaline earth metal complexes were employed as catalysts for amine-silane dehydro coupling reactions. The result showed high reactivity and selectivity of catalyzed silamine and could be isolated up to 23% enantiomeric excess (ee) as determined by chiral HPLC. Hill *et al.* introduced chelating *rac*-DACH-bridged ligands (DACH = *trans*-1,2-diaminocyclohexane) for the synthesis of group 3 and group 13 metal complexes (**H**, Figure 1.10).[31] These bis-NPN aluminium and rare-earth metal complexes showed high activity in the stereoselective polymerization of α-methyl methacrylate. The isolated polymers with the yttrium complex show predominately isotactic (>80 %) close to monodisperse which demonstrates that these complexes are efficient single-site catalytic systems for methyl methacrylate (MMA) polymerization. From the same class of C_2 symmetric ligands, more recently the group of Sundermeyer reported two *N*-aryl bis(iminophosphonamines) based on an enantiomerically pure 1,1'-binaphthyl group.[85] However, the coordination chemistry of type **I** ligands is yet unknown.

1.4 Hydroboration

1.4.1 General

Hydroboration corresponds to the addition of B-H bonds to unsaturated molecules such as aldehydes, ketones, imines, alkenes and alkynes. The first discovery of this reaction was reported by Brown in 1956 in the reaction of alkenes with sodium borohydride-aluminium trichloride.[86] The products obtained from hydroboration reactions can be further hydrolysed, leading to a plethora of useful organic compounds such as alcohols, amines, and alkyl halides. Usually, the hydroboration reaction readily occurs when reactive (electrophilic) boranes, such as diborane (B_2H_6), the borane etherate adduct ($BH_3{\cdot}THF$) or Piers' borane ($HB(C_6F_5)_2$), are added to an alkene. After a suitable workup procedure, the anti-Markovnikov alkene hydration product can be isolated.[87] However, mild reductants, in which the boron atom is bonded to a heteroatom (as in the case of pinacolborane (HBpin) or catecholborane (HBcat)), are active only at elevated temperatures.

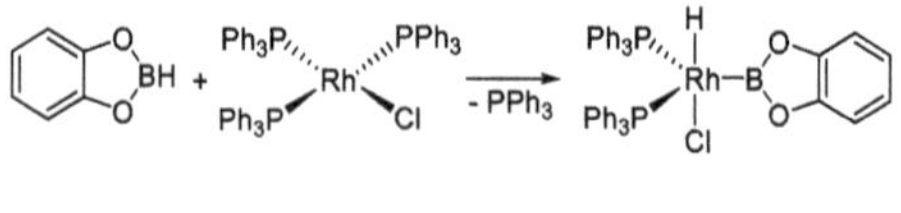

$$\text{HBcat} + \text{RhCl(PPh}_3)_3 \xrightarrow{-\,\text{PPh}_3} \text{RhHCl(Bcat)(PPh}_3)_2$$

Scheme 1.12 Reaction of Wilkinson's catalyst with catecholborane.

Therefore, catalytic hydroboration is desirable to allow this process to occur at ambient conditions. The story of catalytic hydroboration began in 1975 by the observation of Kono and Ito that Wilkinson's catalyst [$Rh(PPh_3)_3Cl$] undergoes oxidative addition upon treatment with catecholborane (Scheme 1.12).[88] The landmark discovery came in 1985, when Männig and Nöth showed that Wilkinson's catalyst can catalyze the hydroboration of alkenes and alkynes.[89] This discovery is the first example of metal-catalyzed hydroboration and further opened the way to other transition metal-catalyzed hydroboration reactions, which is an active area of research.

1.4.2 Hydroboration of carbonyl compounds

The catalytic hydroboration of carbonyl (C=O) compounds emerged as an atom economic route for the synthesis of functionalized alcohols *via* an intermediate borate ester.[90] Traditionally, reduction of carbonyl compounds to alcohols has been performed by treatment with hydrides such as $LiAlH_4$ and $NaBH_4$ which requires appropriate handling due to the reactive nature of the latter reagents. These hydride reagents however can show interaction towards other functional groups that can lead to side reactions.[91] From a green and sustainable chemistry point of view, earth abundant metal complexes have been considered as benign catalysts for hydroboration reactions. In this context, alkaline earth metal-based catalysts are environmentally friendly and cost effective. Furthermore, they present interesting properties such as a low electronegativity, which leads to a high Brønsted basicity responsible for activation of synthetic intermediates. The literature-known alkaline earth metal catalysts used in the hydroboration of a variety of carbonyl compounds are depicted in Figure 1.11.

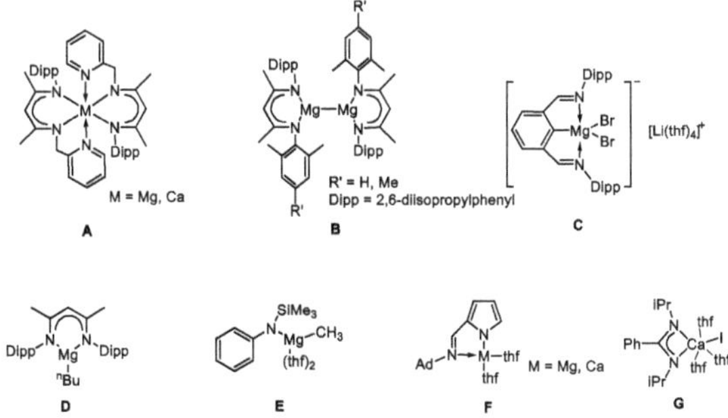

Figure 1.11 Examples of alkaline earth metal catalysts for the hydroboration of carbonyl compounds.[92]

1.5 Photoluminescence

1.5.1 General

Photoluminescence spectroscopy is a non-destructive method to inspect the photophysical properties of a material, also abbreviated as PL. In this process, an electromagnetic wave (photons) is directed to a photoluminescent substance and causes electronic excitation from ground state to excited state. When the substance returns back from the higher state to its ground state, the emission of energy in terms of photons occurs. The emitted light is called luminescence and the process is termed as photoluminescence.

Photoluminescence is further divided into two categories: (i) fluorescence and (ii) phosphorescence. As shown in the Jablonski diagram, when a molecule absorbs a photon of appropriate energy, photoexcitation occurs from the ground singlet state (S_0) into one of the excited singlet states (S_1, S_2, ...) (Figure 1.12). After the photoexcitation, the molecule stays in a nonequilibrium state and rapidly loses energy through vibrational relaxation to attain the vibronic ground state of the corresponding electronically excited state. Further, upon internal conversion, the excited molecule can return back into its lowest excited singlet state (S_1). The radiative decay of the molecule from the S_1 state back to the original S_0 state is an allowed transition (since both states have the same spin multiplicity) resulting in prompt photoluminescence that occurs in the picosecond to nanosecond time scale and is called fluorescence. Alternatively, the molecule can undergo intersystem crossing from the first electronically excited singlet (S_1) to the corresponding triplet state.

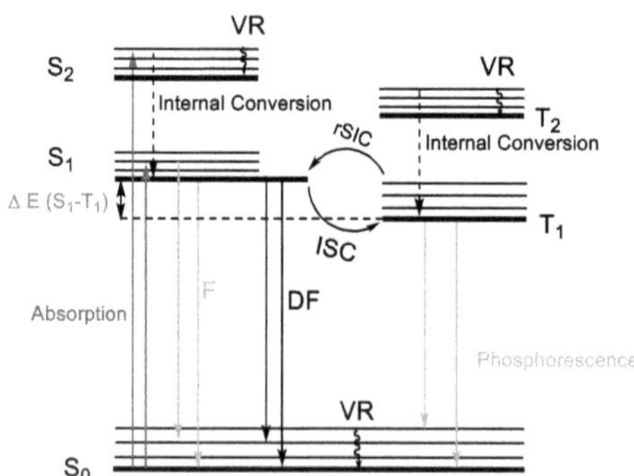

Figure 1.12 Jablonski diagram. S_0: ground singlet state; S_1: first excited singlet state; S_2: second excited singlet state; T_1: first excited triplet state; T_2 second excited triplet state; F: fluorescence; DF: delayed fluorescence; $\Delta E(S_1\text{-}T_1)$: energy gap between S_1 state and T_1 state; VR: vibrational relaxation; ISC: intersystem crossing; rISC: reversible intersystem crossing.

The resulting relaxation of the molecule from T_1 to S_0 is known as phosphorescence. This process is however spin-forbidden and, therefore, phosphorescence occurs with a much longer time scale of micro to milliseconds as compared to fluorescence.[93]

1.5.2 Thermally activated delayed fluorescence (TADF)

When the energy gap between the first excited singlet state S_1 and the first excited triplet state T_1 is small enough (typically $\Delta E_{ST} \leq 0.2$ eV), the molecule can undergo reverse intersystem crossing (Figure 1.12).[94] In this process, the S_1 state is repopulated, which results in a delayed fluorescence (DF). This phenomenon of delayed fluorescence was first reported in uranyl salts by Perrin *et al.* back in 1929.[95] Later in 1941, Lewis *et al.* examined this phenomenon in more detail.[96] The delayed fluorescence was also observed in Eosin, an organic molecule, by Parker *et al.*[97] The photophysical study of the Eosin molecule is responsible for the original name called as E-type delayed fluorescence. In 2012, this phenomenon was rephrased by Adachi and co-workers as thermally activated delayed fluorescence (TADF), which incorporate surrounding thermal energy for the triplet harvesting.[98]

To improve the efficiency of organic light emitting diodes (OLEDs), three generations of OLEDs have been developed. The first-generation OLEDs work on the fluorescence mechanism where efficiency is limited by only 25% singlet excitons based on spin statistics.[99] In order to improve the efficiency of OLEDs the second generation of OLEDs was developed which works on the principle of phosphorescence. The second-generation OLEDs utilize heavy metals such as iridium and platinum that cause spin-orbit coupling and promoting spin forbidden (T_1-S_0) transitions. Although IQEs (IQE = internal quantum efficiency) of 100% may be possibly achieved with the second-generation OLEDs, the costly heavy metals limit their large-scale production. In order to overcome the limitations of the first two generations, a third generation of OLEDs has been developed. In the 3rd-generation OLEDs, the S_1 and T_1 states are tuned in such a way that both of these states are in close proximity to each other, so that the excitons produced in the T_1 state can go through a thermally assisted reverse intersystem crossing into the S_1 state. From this S_1 state, radiative decay into the S_0 state results in delayed fluorescence. Therefore, TADF has become an effective tool to develop cost effective OLEDs.

The TADF concept for effective OLEDs by harvesting the triplet excitons was first established by Adachi *et al.* in 2009 based on their study of a Sn(IV)-porphyrin complex.[100] To get an effective TADF emitter, it is important to have a small energy gap between the singlet and triplet excited states (ΔE_{ST}). By modifying the structural aspects of the molecules, lots of TADF emitters have been reported and are applied industrially to achieve highly efficient OLEDs.[94,101]

2 Aim of the Project

Bearing in mind the intriguing results obtained when using chiral amidinate (NCN) ligands, fine tuning in the ligand backbone could lead to new chiral iminophosphonamide (NPN) ligand systems which await exploration. The objective of this thesis is to get insight into the synthesis and structural features of novel chiral iminophosphonamines and their various metal complexes. To study the coordination behaviour in detail, the new chiral pro-ligands (*R*)-HPEPIA (*P,P*-diphenyl-*N,N'*-bis((*R*)-1-phenylethyl)phosphinimidic amine), (*R*)-HPE^{Dipp}PIA (*P,P*-diphenyl-*N*-((*R*)-1-phenylethyl),*N'*-(2',6'-diisopropylphenyl)phosphinimidic amine) and (*R*)-HNEPIA (*P,P*-diphenyl-*N,N'*-bis((R)-1-naphthylethyl)phosphinimidic amine) bearing one or two stereogenic centres and with different steric bulk are considered (Chart 2.1).

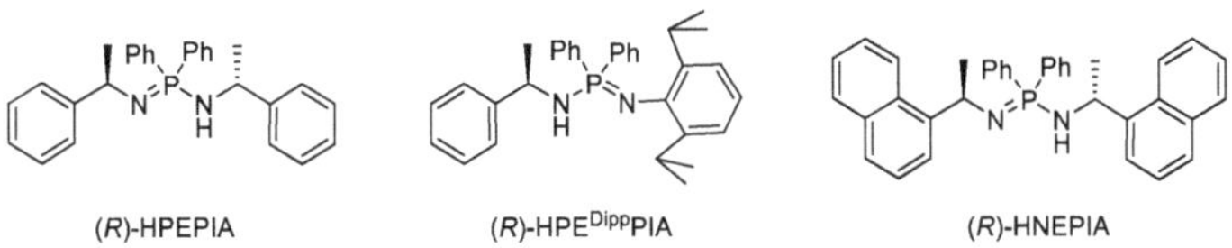

(*R*)-HPEPIA (*R*)-HPE^{Dipp}PIA (*R*)-HNEPIA

Chart 2.1 Chemical structures of chiral iminophosphonamines.

Following are the aims of the project:

1. The first target is to synthesise s-block metal complexes bearing the new iminophosphonamide ligands and to further study their photoluminescence (PL) behaviour. Previous results in the group have indeed highlighted the high luminescence of earth abundant metal complexes with diamidophosphine ($PNNP^{2-}$) ligands.[102] The synthesised alkaline earth metal complexes will be tested in catalysis owing to their relatively cheap and environmentally friendly nature.
2. The PL studies will be further extended to Cu(I) and Zn(II) complexes using the above-mentioned ligands (Chart 2.1).
3. Further, the chiral iminophosphonamides will be introduced to group 13 metal centres with the aim to get complexes active in Lewis acid mediated catalysis.
4. Inspired by the catalytic activity of lanthanide complexes based on chiral amidinate (*S*)-HPEBA ligands (Figure 1.9 (D)), the introduction of chiral iminophosphonamines in the coordination sphere of lanthanides will be also studied.

5. Finally, the reactivity of chiral iminophosphonamides and chiral amidinates will be tested with low-valent group-14 compounds.

3 Results and Discussion

3.1 Introduction

The monoanionic iminophosphonamides, having the general formula $[(R_2P(NR')(NR'')]^-$, are generated by a simple deprotonation from their corresponding iminophosphonamines. These ligands coordinate mostly in chelating modes as described in section 1.2. Although achiral versions of iminophosphonamines are known with many differently substituted NPN backbones, the coordination chemistry of their corresponding chiral congeners is limited only to a few ligand backbones, as already discussed in section 1.3. The steric and electronic influence of the substituents affect the different binding modes. In this regard, by changing the substituents on the two nitrogen atoms, a variety of symmetrical and unsymmetrical iminophosphonamines can be synthesized. In the present work, three selected differently substituted chiral iminophosphonamines (Chart 3.1) and their coordination chemistry have been investigated.

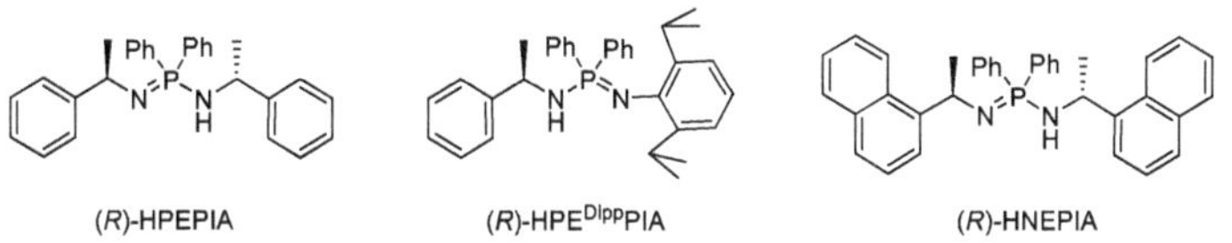

(*R*)-HPEPIA (*R*)-HPE^{Dipp}PIA (*R*)-HNEPIA

Chart 3.1 The chiral iminophosphonamines (*R*)-HPEPIA, (*R*)-HPE^{Dipp}PIA and (*R*)-HNEPIA.

The synthesis and structural characterization of the symmetric ligand *P,P*-diphenyl-*N,N'*-bis((*R*)-1-phenylethyl)phosphinimidic amine ((*R*)-HPEPIA) (Scheme 3.1) has already been published in the collaborative work with Dr. Thomas J. Feuerstein and also been discussed in his dissertation.[103]

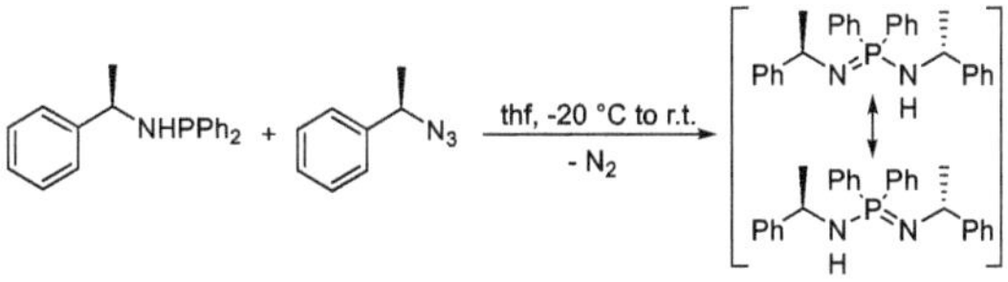

NHPPh$_2$ + N$_3$ thf, -20 °C to r.t. - N$_2$

Scheme 3.1 Synthesis of (*R*)-HPEPIA.

3.1.1 Synthesis of *P,P*-diphenyl-*N*-((*R*)-1-phenylethyl),*N'*-(2',6'-diisopropylphenyl)phosphinimidic amine, ((*R*)-HPE^{Dipp}PIA)

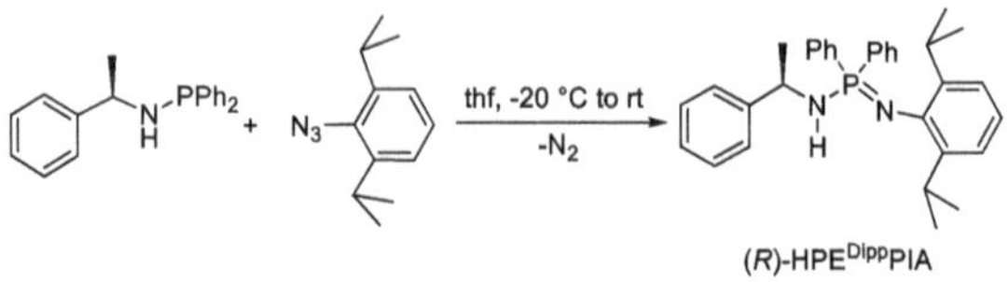

thf, -20 °C to rt
$-N_2$
(*R*)-HPE^{Dipp}PIA

Scheme 3.1.1 Synthesis of (*R*)-HPE^{Dipp}PIA.

Following a similar synthetic route as used for (*R*)-HPEPIA and other closely related ligands, the desired unsymmetrical iminophosphonamine ligand (*R*)-HPE^{Dipp}PIA was prepared *via* a Staudinger reaction between Dipp-N_3[104] (Dipp = 2,6-$^{i}Pr_2C_6H_3$) and HN(*R*-CH(CH_3)Ph)(PPh_2)[80c,105] in a straightforward manner (Scheme 3.1.1).[25a,26,103b]

The ligand was further characterized by several analytical methods such as multinuclear NMR, IR, elemental analysis and single crystal X-ray diffraction. The characteristic septet of the C*H*(CH_3)$_2$ protons is observed at δ = 3.67 ppm, whereas the corresponding resonance for the methine proton attached to chiral center (*i.e.* Ph(C*H*)CH_3) is observed at δ = 4.64 ppm in the ^{1}H NMR spectrum. The resonances for Ph(CH)C*H*$_3$ and CH(C*H*$_3$)$_2$ are detected at δ = 1.26 ppm and at δ = 1.18 ppm, respectively. In contrast to the broad resonance observed for the N*H* proton in (*R*)-HPEPIA, a distinct apparent triplet at δ = 2.81 ppm ($^2J_{HH}$ = 9.5 Hz, $^2J_{PH}$ = 9.5 Hz) could be assigned to the N*H* proton for (*R*)-HPE^{Dipp}PIA. This indicates that (*R*)-HPE^{Dipp}PIA does not involve into any intramolecular N-H···N exchange in difference to (*R*)-HPEPIA, as observed in previous studies.[103b] The ^{31}P{^{1}H} NMR spectrum of (*R*)-HPE^{Dipp}PIA exhibits a single signal at δ = -10.9 ppm, which is significantly shifted up field in comparison to that of the parent compound HN(*R*-CH(CH_3)Ph)(PPh_2) (δ = 36.6 ppm). Furthermore, the N-H stretching mode of (*R*)-HPE^{Dipp}PIA in the IR spectrum is detected at 3347 cm^{-1}.

Single crystals of (*R*)-HPE^{Dipp}PIA suitable for single crystal X-ray diffraction were grown from hot *n*-heptane. It crystallized in the orthorhombic chiral space group $P2_12_12_1$ with one molecule of the ligand in the asymmetric unit (Figure 3.1.1). The phosphorous atom of the N-P-N moiety is tetrahedrally coordinated. The P-N1 (1.536(4) Å) bond is significantly shorter than the P-N2 (1.660(4) Å) bond, which suggests the presence of a double bond localized between P-N1. The hydrogen atom could be refined in the difference Fourier map and can be assigned to N2.[103b] The N1-P-N2 bond angle (116.7(2)°) is smaller than in the symmetrical (*R*)-HPEPIA ligand and comparable to similar achiral compounds.[25c,80c,103b]

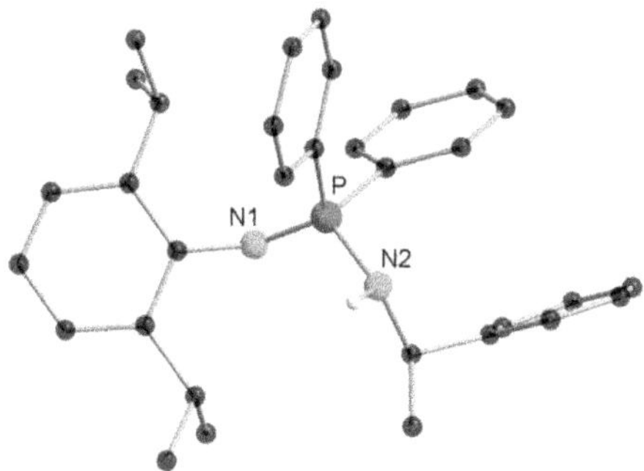

Figure 3.1.1 Molecular structure of (*R*)-HPE^{Dipp}PIA in the solid state. All hydrogen atoms except the amine proton are omitted for clarity. Selected bond lengths (Å) and bond angles [°]: P-N1 1.536(4), P-N2 1.660(4); N1-P-N2 116.7(2).

3.1.2 Synthesis of *P,P*-diphenyl-*N,N'*-bis((R)-1-naphthylethyl)phosphinimidic amine ((*R*)-HNEPIA)

The symmetrically substituted (*R*)-HNEPIA ligand was synthesized following the similar synthetic route as described for (*R*)-HPEPIA and (*R*)-HPE^{Dipp}PIA. The Staudinger reaction between HN((*R*)-CH(CH_3)naphPPh$_2$),[106] and (*R*)-CH(CH_3)naph)N_3,[107] (naph = naphthyl) afforded (*R*)-HNEPIA as a white solid (Scheme 3.1.2).

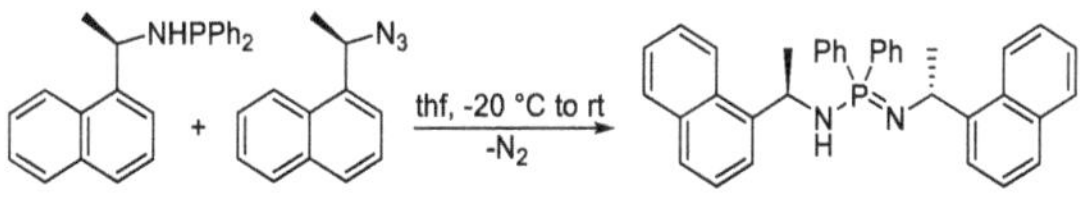

NHPPh$_2$ + N$_3$ → (thf, -20 °C to rt, -N$_2$) → (*R*)-HNEPIA

Scheme 3.1.2 Synthesis of (*R*)-HNEPIA.

The ^{1}H NMR spectrum of the ligand exhibit different signals for methine and methyl protons, which in consistence with previously reported (*R*)-HPEPIA ligand.[103b] These resonances exhibits two sets of signals for naph(CH)C*H*$_3$ (δ = 1.81, 1.21 ppm) (naph = naphthyl) and naph(C*H*)CH$_3$ (δ = 5.45-5.52, 5.17 ppm) in the ^{1}H NMR spectrum. The corresponding N*H* proton is observed as a broad signal at δ = 2.88 ppm. Furthermore, the non-equivalent nature of the methine and methyl carbons of the (*R*)-HNEPIA ligand is also evidenced by the appearance of two sets of signals in pairs for naph(CH)*C*H$_3$ (δ = 29.9, 25.7 ppm) and naph(*C*H)CH$_3$ (δ = 51.2, 46.3 ppm) in the $^{13}C\{^1H\}$ NMR spectrum. The purity of the ligand is further confirmed by analysis of the $^{31}P\{^1H\}$ NMR spectrum, where only one single resonance is observed at δ = 3.6 ppm. In the IR spectrum, a band at 3372 cm^{-1} is assigned to the $\tilde{\nu}_{NH}$ stretching mode.

The protonated ligand (*R*)-HNEPIA was crystallized from a hot *n*-heptane solution. It crystallizes in the monoclinic chiral space group $P2_1$ with two molecules in the asymmetric unit (Figure 3.1.2). The P-N1 bond length (1.675(4) Å) is within the range of a P-N single bond,[103b] whereas the P-N2 bond length of 1.544(4) Å is in the range of a P-N double bond. The hydrogen atom could be freely refined and assigned to N1. The N1-P-N2 bond angle (119.6(2)°) is wider than that observed in (*R*)-HPE^{Dipp}PIA (116.7(2)°), but narrower than that in (*R*)-HPEPIA (121.6(2)°).

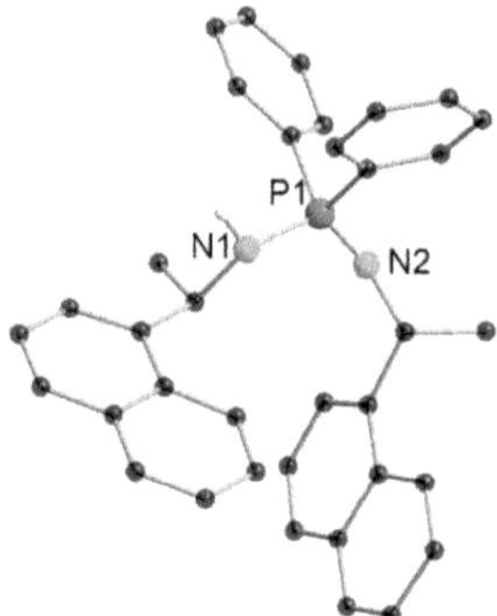

Figure 3.1.2 Molecular structure of ((*R*)-HNEPIA) in the solid state. All hydrogen atoms except the amine proton are omitted for clarity. Selected bond lengths (Å) and bond angles [°]: P1-N1 1.675(4), P1-N2 1.544(4); N1-P1-N2 119.6(2).

3.1.3 Synthesis of the enantiopure alkali metal complexes of iminophosphonamide

3.1.3.1 Synthesis of [M_2{(*R*)-PEPIA}$_2$] (M = K (**1**), Rb (**2**), and Cs (**3**))

The results discussed in this section and section 3.1.4 have already been partly published in:

T. J. Feuerstein, B. Goswami, P. Rauthe, R. Köppe, S. Lebedkin, M. M. Kappes, P. W. Roesky, Alkali metal complexes of an enantiopure iminophosphonamide ligand with bright delayed fluorescence, *Chem. Sci.*, **2019**, 10, 4742–4749. In this contribution, the synthesis of M_2{(*R*)-PEPIA}$_2$ (M = Li and Na) has already been discussed in the dissertation of Dr. Thomas Feuerstein.[103a] Herein, the synthesis and characterization of the corresponding potassium, rubidium and cesium complexes are described.

Alkali metal complexes are the most common precursors to transfer ligands via salt metathesis reactions on different elements across the periodic table. Interestingly, in many cases, the choice of the alkali metal salt is crucial to obtain a specific desired product. In general, lithium salts have the advantage of better solubility in organic solvents. In contrast, the potassium salts, when reacted with corresponding chloride precursors, lead to the formation of the byproduct KCl, which is insoluble in most organic solvents and can be separated easily.

In this regard, the preparation of a library of alkali metal complexes is of special interest. The deprotonation of (*R*)-HPEPIA by KH in THF resulted in the formation of the dimeric complex [{(*R*)-PEPIA}$_2K_2$] (**1**) (Scheme 3.1.3). The identity of complex **1** was confirmed by using various analytical techniques such as ^{1}H, ^{13}C{^{1}H}, and ^{31}P{^{1}H} NMR, IR, single crystal X-ray diffraction and elemental analysis.

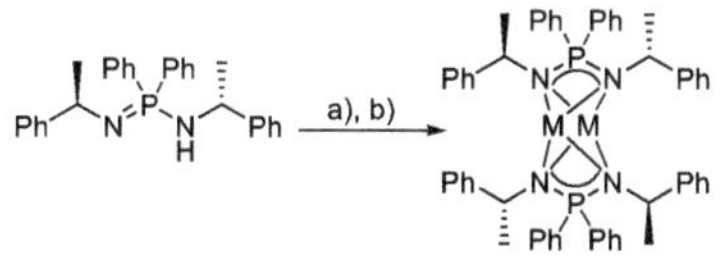

M = K (**1**), Rb (**2**)

Scheme 3.1.3 Synthesis of dimeric [{(*R*)-PEPIA}$_2K_2$] (**1**), and [{(*R*)-PEPIA}$_2Rb_2$] (**2**) with following synthetic routes: a) KH, THF, r.t., 16 h, b) $RbN(SiMe_3)_2$, toluene, r.t., 3 d; or Rb, toluene, 90 °C to r.t., 3 h.

The ^{1}H NMR spectrum of **1** shows a single set of signals for the different substituents, suggesting the symmetric nature of the complex. The methine protons Ph(C*H*)CH_3 appear as a doublet ($^3J_{PH}$ = 23.4 Hz) of quartets ($^3J_{HH}$ = 6.3 Hz) at δ = 4.05 ppm and the methyl protons appear as a doublet at δ = 0.92 ppm ($^3J_{HH}$ = 6.4 Hz). These resonances are slightly shifted upfield in comparison to those in its starting precursor, (*R*)-HPEPIA.[103b] Complex **1** exhibits a single resonance at δ = 16.3 ppm in the $^{31}P\{^1H\}$ NMR spectrum, which is significantly shifted downfield as compared to that in the parent compound (*R*)-HPEPIA (δ = 2.7 ppm). Disappearance of the N-H stretching band in the IR spectrum of **1** further supports the complete deprotonation of the ligand. Colourless crystals of **1** suitable for X-ray diffraction analysis were obtained from a hot toluene solution. Complex **1** crystallizes as a dimer in the orthorhombic chiral space group $P2_12_12_1$ with one molecule in the asymmetric unit (Figure 3.1.3).

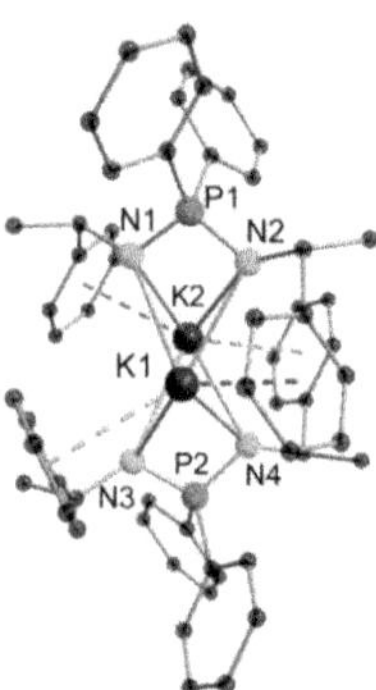

Figure 3.1.3 Molecular structure of **1** in the solid state. All of the hydrogen atoms are omitted for clarity. Selected bond lengths [Å] and angles [°]: K1-N1 2.754(2), K1-N2 3.043(2), K1-N3 2.802(2), K1-N4 3.130(2), K2-N2 2.878(2), K2-N3 3.115(2), K2-N4 2.733(2), P1-N1 1.591(2), P1-N2 1.602(2), P2-N3 1.598(2), P2-N4 1.591(2); N1-K1-N3 128.06(6), N1-K1-N4 93.32(6), N2-K1-N3 94.82(6), N2-K1-N4 102.11(6), N1-K1-N2 52.57(6), N3-K1-N4 50.65(5), N2-K2-N3 91.78(6), N2-K2-N4 117.69(6), N3-K2-N4 51.29(6), K1-N2-K2 69.58(5), K1-N3-K2 69.46(5), K1-N4-K2 70.06(5), N1-P1-N2 107.82(11), N3-P2-N4 106.47(10).Torsion angle [°]: N1-P1-P2-N3 146.510(12). K···K and P···P distance [Å]: K1···K2 3.3809(6), P1-P2 6.1055(10). M···C coordination interactions [Å]: K1···C3 3.100(2), K1···C4 3.408(3), K1···C7 3.488(3), K1···C8 3.162(3),K1···C19 3.151(2), K1···C20 3.147(3), K1···C21 3.517(3), K1···C24 3.517(3), K2···C11 3.044(2), K2···C12 3.220(3), K2···C13 3.528(3), K2···C15 3.497(3), K2···C16 3.196(3), K2···C27 3.177(2), K2···C28 3.373(3), K2···C32 3.387(3).

In contrast to dimeric nature of complex **1**, the corresponding potassium iminophosphonamide compounds $[K\{Ph_2P(Me_3SiN)_2\}(thf)_4]$, $[K\{(2,6\text{-}i\text{Pr}_2C_6H_3N)_2P(Ph_2)\}]_n$[26] and $[K\{(2,4,6\text{-}Me_3C_6H_2N)_2P(Ph_2)\}]_n$[29] are monomeric or polymeric, respectively. It is thus anticipated that the substituents present on the nitrogen atoms play a crucial role in the aggregation state of the potassium compounds. The solid-state structure of **1** suggests, that the potassium centres bind in bridging as well as in chelating fashions with K1, K2 (μ- κ^2: κ^2) binding mode. The average K-N bond length observed in **1** is 2.921 (2) Å, which is similar to that in $[K\{Ph_2P(Me_3SiN)_2\}(thf)_4]$.[32] The P-N bond length is in the range between a single and a double bond, suggesting a delocalisation of the negative charge over the N-P-N moiety. The average N-P-N bond angle is 107.14(10)°. Due to the larger ionic radius of potassium, the torsion angle N1-P1-P2-N3 for **1** (146.510(12)°) is wider than those in the corresponding lithium (88.6(1)°) and sodium (108.7(3)°) analogues already described.[103b] The molecular structure in the solid state of complex **1** shows an intermolecular interaction between potassium atoms and phenyl rings with a K···C interaction in the range of 3.044(2)-3.528(3) Å,[108] out of which the individual potassium centres show two different hapticities, (η^4, η^4) for K1 and (η^5, η^3) for K2, with the phenyl rings.

In order to understand the coordination behaviour of {(*R*)-PEPIA}$^-$, the heavier alkali metal rubidium was utilized. Treatment of (*R*)-HPEPIA with elemental Rb or $Rb(N(SiMe_3)_2)$ yielded the dimeric complex [{(*R*)-PEPIA}$_2$Rb$_2$] (**2**) (Scheme 3.1.3, M = Rb). The $^{31}P\{^1H\}$ NMR spectrum of complex **2** shows a single resonance at δ = 14.7 ppm, which is shifted downfield compared to that in (*R*)-HPEPIA (δ = 2.7 ppm), suggesting a complete deprotonation of the ligand. The absence of an N-H stretching band in the IR spectrum of complex **2** and the disappearance of the N-*H* resonance in the ^1H NMR spectrum further confirms the deprotonation of the ligand by Rb or $RbN(SiMe_3)_2$. The appearance of a single set of signals in the ^1H NMR spectrum indicates a symmetric coordination of ligand in solution. A characteristic doublet corresponding to the Ph(CH)C*H*$_3$ protons at δ = 0.96 ppm with a coupling constant of $^3J_{HH}$ = 6.4 Hz can be observed. However, the corresponding Ph(CH)C*H*$_3$ resonances for the protonated ligand (*R*)-HPEPIA were observed at δ = 1.66 and δ = 1.13 ppm due to proton (N*H*) exchange (tautomerization). A doublet of quartets at δ = 4.06 ppm ($^3J_{PH}$ = 23.4 Hz, $^3J_{HH}$ = 6.4 Hz, respectively) is assigned to the methine protons (*i.e.* Ph(C*H*)CH$_3$) of complex **2**. In addition, due to the interaction between the aromatic rings and the rubidium atoms, the aromatic resonances δ = 7.54-6.91 ppm are slightly shifted upfield compared to those observed in the starting precursor (*R*)-HPEPIA (δ = 8.04-6.80 ppm).

Single crystals suitable for X-ray diffraction analysis were obtained from hot *n*-hexane. Complex **2** is isostructural to **1** and crystallizes in the orthorhombic system with the chiral space group $P2_12_12_1$ (Figure 3.1.4). Similar to **1**, complex **2** also shows both chelating and bridging bonding mode for the ligand, (μ- κ^2: κ^2) with Rb1 and (μ- κ^2: κ^1) with Rb2, respectively. The average M–N bond length in **2** (3.0582(3) Å) is larger than that in **1** (2.921(2)

Å) and similar to the Rb-N contact in $[Rb_2\{Ph_2P(Me_3SiN)_2\}_2(thf)_2]$ (2.953(2) Å).[32] Obviously, the larger M–N bond in complex **2** is due to the larger ionic radius of rubidium compared to that of potassium.[109] The individual rubidium centres show two different hapticities, (η^5, η^4) for Rb1 and (η^6, η^4) for Rb2 with the phenyl rings of the ligand back bone with Rb···C interactions in the range of 3.137(4)-3.687(5) Å.[110] However, this kind of Rb-phenyl interaction was not observed in case of $[Rb_2\{Ph_2P(Me_3SiN)_2\}_2(thf)_2]$.

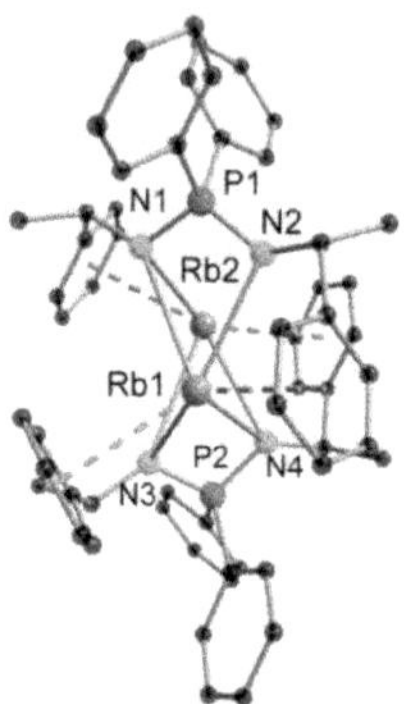

Figure 3.1.4 Molecular structure of **2** in the solid state. Hydrogen atoms are omitted for clarity. Selected bond lengths [Å] and angles [°]: Rb1-N1 2.885(3), Rb1-N2 3.169(3), Rb1-N3 2.936(3), Rb1-N4 3.243(3), Rb2-N2 3.009(3), Rb2-N3 3.250(3), Rb2-N4 2.869(3), P1-N1 1.592(3), P1-N2 1.584(3), P2-N3 1.575(3), P2-N4 1.594(3); N1-Rb1-N3 125.70(9), N1-Rb1-N4 93.96(9), N2-Rb1-N3 94.79(9) N2-Rb1-N4 101.76(8), N1-Rb1-N2 50.05(9), N3-Rb1-N4 48.59(8), N2-Rb2-N3 91.77(8), N2-Rb2-N4 115.61(9), N3-Rb2-N4 48.93(8), Rb1-N2-Rb2 70.77(7), Rb1-N3-Rb2 70.52(6), Rb1-N4-Rb2 71.42(7), N1-P1-N2 108.5(2), N3-P2-N4 107.2(2). Torsion angle [°]: N1-P1-P2-N3 147.711(13). Rb2-N1, M-M and P-P distances [Å]: Rb2-N1 3.6114(5), Rb1···Rb2 3.5804(6), P1-P2 6.2784(11). Rb···C coordination interactions [Å]: Rb1 ··C3 3.171(4), Rb1···C4 3.408(4), Rb1···C5 3.687(5), Rb1···C7 3.509(5), Rb1···C8 3.245(4) Rb1···C19 3.206(4), Rb1···C20 3.208(4), Rb1···C21 3.486(4), Rb1···C24 3.492(4), Rb2···C11 3.137(4), Rb2···C12 3.279(4), Rb2···C13 3.554(5), Rb2···C14 3.685(5), Rb2···C15 3.555(5), Rb2···C16 3.289(4) Rb2···C27 3.230(4), Rb2···C28 3.373(4), Rb2···C29 3.666(5), Rb2···C32 3.411(4).

Stalke *et al.* reported that the reaction of elemental caesium with the iminophosphonamine $Ph_2P(Me_3SiN)_2H$ leads to the formation of a dimeric complex $[Ph_2P(Me_3SiN)_2Cs]_2$, which further exists in a polymeric form when considering the very close intramolecular metal-phenyl interactions.[32] To examine whether a similar coordination behaviour is realized for $\{(R)\text{-PEPIA}\}^-$, the equimolar reaction of (*R*)-HPEPIA with either elemental Cs or $CsN(SiMe_3)_2$ was carried out. Both routes led to the formation of complex **3** (Scheme 3.1.4).

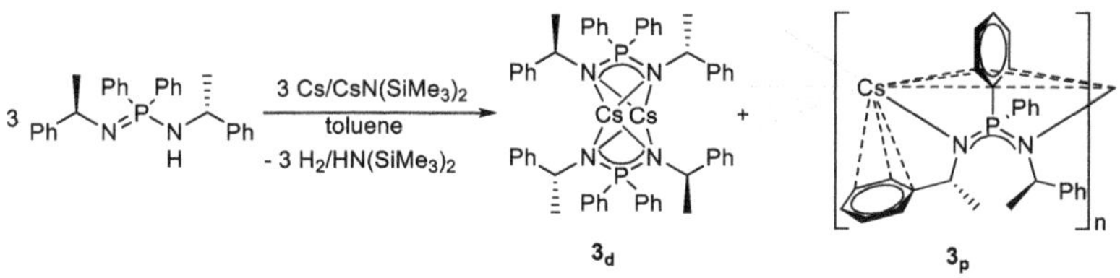

Scheme 3.1.4 Synthesis of [{(*R*)-PEPIA}$_2$Cs$_2$]/[{(*R*)-PEPIA}Cs]$_n$ (**3**)

Crystals suitable for X-ray diffraction analysis were obtained from a hot *n*-hexane solution. Complex **3** crystallized in the orthorhombic lattice with the chiral space group $P2_12_12_1$. The solid-state structure of **3** shows two different coordination mode of caesium with {(*R*)-PEPIA}$^-$, likely due to larger ionic radius of caesium. Surprisingly, complex **3** crystallizes as a cocrystal with a 1 : 1 ratio of the dimer [{(*R*)-PEPIA}$_2$Cs$_2$] (**3$_d$**) and the coordination polymer [{(*R*)-PEPIA}Cs]$_n$ (**3$_p$**) in the asymmetric unit (Figure 3.1.5).

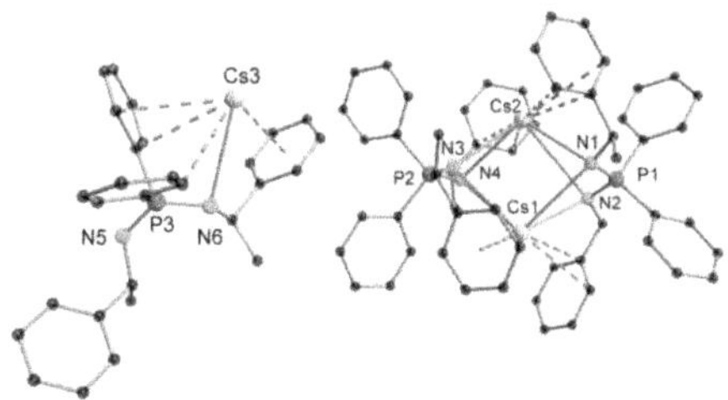

Figure 3.1.5 Full asymmetric unit of **3** in the solid state. Hydrogen atoms are omitted for clarity.

Similar to complex **1** and **2**, the dimeric form of **3$_d$** also shows chelating as well as bridging modes for the ligand (*R*)-HPEPIA (Figure 3.1.6). It could be anticipated that with increasing the ionic radius of the central metal atoms, the N1-P1-P2-N3 torsion angles increase from 146.510(12)° (**1**), to 147.711(13)° (**2**) and finally to 163.23(2)° (**3$_d$**), resulting in a nearly planar arrangement of the ligands in **3$_d$.**

Throughout the series of the dimeric alkali metal complexes, [{(*R*)-PEPIA}$_2$M$_2$] {M = Li, Na, K (**1**), Rb (**2**), Cs (**3$_d$**)}, it is estimated that the bonding mode and orientation of the ligand depends on the ionic radius of the respective central metal atom.[103b] This size difference eventually alters the M···M distances in the corresponding complexes (Table 3.1.1).

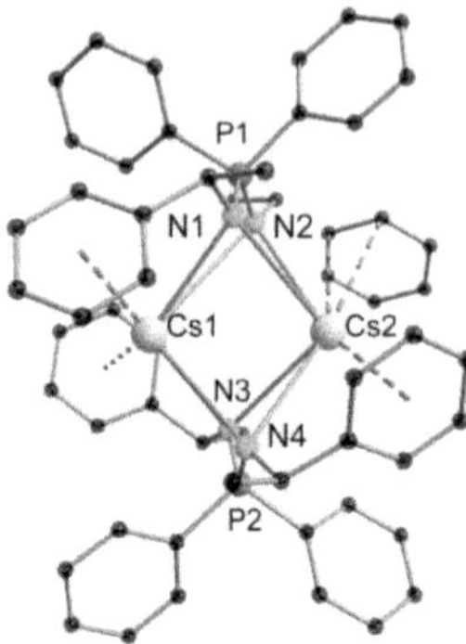

Figure 3.1.6 Molecular structure of $\mathbf{3_d}$ in the solid state. Hydrogen atoms are omitted for clarity. Selected bond lengths [Å] and angles [°]: Cs1-N1 3.114(6), Cs1-N2 3.703(6), Cs1-N3 3.207(6), Cs1-N4 3.199(6), Cs2-N1 3.438(6), Cs2-N2 3.053(6), Cs2-N3 3.289(6), Cs2-N4 3.251(6), P1-N1 1.589(6), P1-N2 1.583(6), P2-N3 1.605(6), P2-N4 1.600(6); N1-Cs1-N3 111.0(2), N1-Cs1-N4 95.2(2), N2-Cs1-N3 83.4(2), N2-Cs1-N4 100.58(14), N1-Cs1-N2 43.25(14), N3-Cs1-N4 47.1(2), N1-Cs2-N3 101.5(2), N1-Cs2-N4 88.3(2), N2-Cs2-N3 93.3(2), N2-Cs2-N4 115.0(2), N1-Cs2-N2 46.2(2), N3-Cs2-N4 46.1(2), Cs1-N1-Cs2 72.95(13), Cs1-N2-Cs2 69.85(12), Cs1-N3-Cs2 73.86(13), Cs1-N4-Cs2 74.48(13), N1-P1-N2 108.3(3), N3-P2-N4 106.3(3). Torsion angle [°]: N1-P1-P2-N3 163.23(2). M-M and P-P distance [Å]: Cs1···Cs2 3.9038(8), P1-P2 6.498(3). Cs1···C3 3.467(7), Cs1···C4 3.563(9), Cs1···C8 3.762(9), M···C coordination interactions [Å]: Cs1···C27 3.574(7), Cs1···C31 3.700(7), Cs1···C32 3.402(7), Cs2···C11 3.531(7), Cs2···C16 3.568(8), Cs2···C19 3.481(8), Cs2···C20 3.554(9), Cs2···C21 3.700(9), Cs2···C22 3.730(8), Cs2···C23 3.730(8), Cs2···C24 3.580(9).

Besides, as the size of the central metal increases, the M···C interactions increase from [{(*R*)-PEPIA}$_2$Li$_2$] to [{(*R*)-PEPIA}$_2$Rb$_2$] (**2**), however drop for $\mathbf{3_d}$ possibly due to steric considerations.

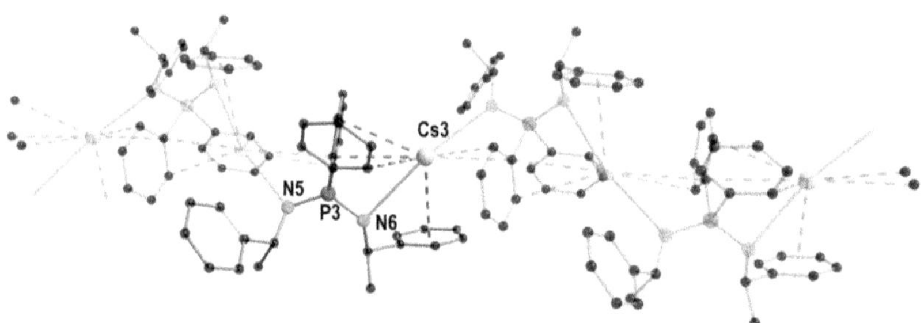

Figure 3.1.7 Asymmetric unit of the cocrystallized coordination polymer $\mathbf{3_p}$. Hydrogen atoms are omitted for clarity. Selected bond lengths [Å] and angles [°]: Cs3-N5 3.104(7), Cs3-N6 3.282(7), P3-N5 1.602(7), P3-N6 1.590(7), N5-Cs3-N6 166.4(2), P3-N6-Cs3 97.8(3), P3-N5-Cs3 114.3(3), N5-P3-N6 124.5(4). M···C coordination interactions [Å]: Cs3···C67 3.546(8), Cs3···C71 3.673(9), Cs3···C72 3.332(10), Cs3···C73 3.459(9), Cs3···C74 3.837(9), Cs3···C78 3.590(10) Cs3···C73 3.476(9), Cs3···C74' 3.526(9), Cs3···C80 3.632(8).

Moreover, the molecular structure of $\mathbf{3_p}$ shows a (μ: κ^1: κ^1) coordination mode with the ligand (Figure 3.1.7). Each caesium atom (Cs3) is almost linearly coordinated by the two nitrogen atoms (N5 and N6) with a bond angle of N5-Cs3-N6 166.4(2)° and bond lengths of 3.104(7) Å (Cs3-N5) and 3.282(7) Å (Cs3-N6).

Table 3.1.1 Selected structural details of alkali metal complexes of (*R*)-HPEPIA.

Complex	M1-M2 distance (in Å)	Coordination mode	N-P-N-angles [°]	N-P-P-N-Torsion angle [°]
[Li_2{(*R*)-PEPIA}$_2$]	2.4094(8)	Li1: η^2; Li2: η^2	N1-P1-N2 108.52(14) N3-P2-N4 99.7(2)	88.6(1)
[Na_2{(*R*)-PEPIA}$_2$]	2.907(4)	Na1: η^3; Na2: η^3	N1-P1-N2 109.6(2) N3-P2-N4 102.8(2)	108.7(3)
1	3.3809(6)	K1: η^4:η^4; K2:η^5:η^3	N1-P1-N2 107.82(11) N3-P2-N4 106.47(10)	146.510(12)
2	3.5804(6)	Rb1: η^5:η^4; Rb2: η^6:η^4	N1-P1-N2 108.5(2) N3-P2-N4 107.2(2)	147.711(13)
$3_{d,p}$	3.9038(8)	**3_d**: Cs1: η^3:η^3; Cs2:η^2:η^6 **3_p**: Cs3: η^3:η^3:η^2:η^1	N1-P1-N2 108.3(3) N3-P2-N4 106.3(2) N5-P3-N6 124.5(4)	163.23(2)

In addition, Cs3 coordinates with phenyl rings *via* an η^3:η^3:η^2:η^1 coordination mode. The Cs-N bond lengths in $\mathbf{3_p}$ are in the close range of those observed for [Cs_2{$Ph_2P(Me_3SiN)_2$}$_2$]$_n$ (3.010(2)-3.14(2) Å).[32] In the ^{1}H NMR spectrum of complex **3**, only one set of signals is observed, indicating its symmetric nature in solution. Appearance of a single peak in the ^{31}P{^{1}H} NMR at δ= 12.6 ppm further confirms the purity of the product.

3.1.3.2 Synthesis of [{(*R*)-PE^{Dipp}PIA}$_2$M$_2$] (M = Li (4), Na (5))

In order to understand the coordination behaviour with the sterically bulkier iminophosphonamine (*R*)-HPE^{Dipp}PIA, an equimolar reaction between (*R*)-HPE^{Dipp}PIA and LiN(SiMe$_3$)$_2$ was carried out in toluene, which resulted in the formation of the lithium complex [{(*R*)-PE^{Dipp}PIA}$_2$Li$_2$] (**4**) (Scheme 3.1.5).

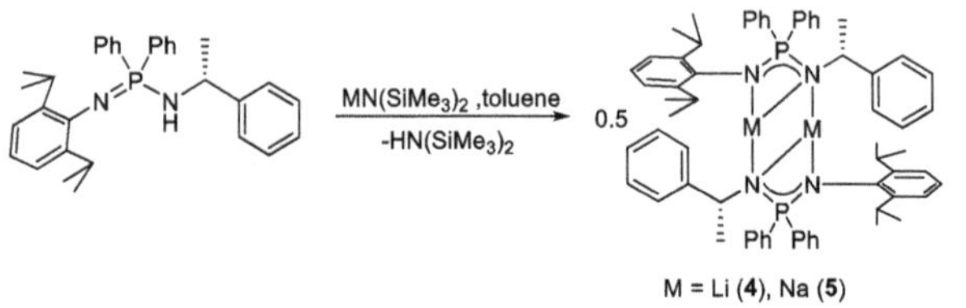

MN(SiMe$_3$)$_2$,toluene
-HN(SiMe$_3$)$_2$
0.5
M = Li (**4**), Na (**5**)

Scheme 3.1.5 Synthesis of [{(*R*)-PE^{Dipp}PIA}$_2$Li$_2$] (**4**) and [{(*R*)-PE^{Dipp}PIA}$_2$Na$_2$] (**5**).

The lithium ions were detected in the ^{7}Li{^{1}H} NMR spectrum as a singlet at δ = 2.9 ppm. The identity of complex **4** was further confirmed by analysis of its ^{1}H NMR spectrum, in which a doublet at δ = 1.75 ppm ($^3J_{HH}$ = 6.64) is assigned to Ph(CH)C*H*$_3$, whereas the Ph(C*H*)CH$_3$ proton appears as a broad signal at δ= 3.12 ppm ($\Delta\nu_{1/2} \approx$ 124.9 Hz). The ^{31}P{^{1}H} NMR spectrum of complex **4**, exhibits a single peak at δ = 28.8 ppm which is significantly shifted downfield compared to that of the starting precursor (*R*)-HPE^{Dipp}PIA (δ= -10.9 ppm).

The molecular structure of complex **4** was determined by single crystal X-ray diffraction analysis. Complex **4** crystallizes in orthorhombic system with the chiral space group $P2_12_12_1$ with one molecule in the asymmetric unit (Figure 3.1.8). Similar to [{(*R*)-PEPIA}$_2$Li$_2$],[103b] complex **4** also exists in a dimeric form, which is in contrast to the monomeric form observed for [{Ph$_2$P(Me$_3$SiN)$_2$}Li(thf)$_2$].[32] Complex **4** could be best described as three four-membered rings fused into a stair-shaped structure. In order to minimise the steric repulsion, the Dipp substituents on the nitrogen atoms are arranged in a *trans* fashion. The two lithium centres in the complex are not coplanar with the N$_2$P rings, instead, they are slightly above the planes of the individual N$_2$P rings with dihedral angles of 12.05(2)° and 18.80(3)° between the individual N$_2$P and N$_2$Li rings.

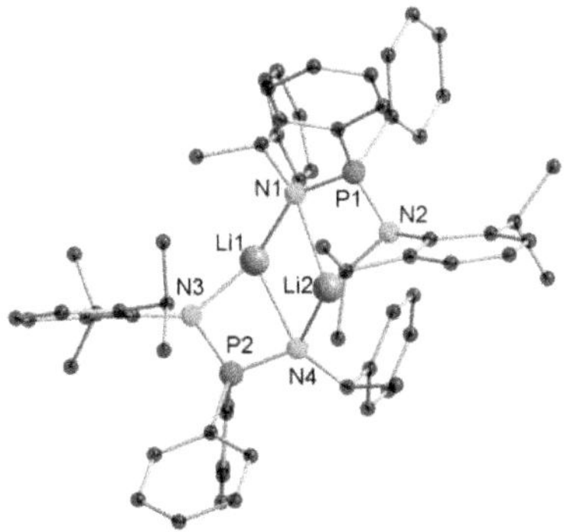

Figure 3.1.8 Molecular structure of **4** in the solid state. Hydrogen atoms are omitted for clarity. Selected bond lengths [Å] and angles [°]: Li1-N1 1.975(9), Li1-N3 1.931(9), Li1-N4 2.209(9), Li2-N1 2.303(9), Li2-N2 1.930(9), Li2-N4 1.955(9), P1-N1 1.614(4), P1-N2 1.596(4), P2-N3 1.594(4), P2-N4 1.618(4); N1-Li1-N3 140.7(6), N3-Li1-N4 75.5(3), N1-Li1-N4 113.0(4), N2-Li2-N4 148.7(6), N1-Li2-N2 73.7(3), N1-Li2-N4 109.9(4), Li1-N4-Li2 68.6(4), N1-P1-N2 105.6(2), N3-P2-N4 104.7(2). Torsion angle [°]: N1-P1-P2-N4 143.23(2). M-M and P-P distance [Å]: Li1-Li2 2.358(2), P1-P2 5.51(2).

In contrast to the dimeric alkali metal complexes, [{(*R*)-PEPIA}$_2$M$_2$] (M = Li, Na, K (**1**), Rb (**2**), Cs (**3d**)), no M-arene interaction is observed for **4**. The P-N bond lengths (1.594(4)-1.618(4) Å) in **4** fall in between the distances reported for P-N single and double bonds, indicating the delocalisation of the negative charge over the N_2P moiety. The Li-N bond length of 2.0505(9) Å in **4** is comparable with the Li-N contact in [Li{Ph$_2$P(Me$_3$SiN)$_2$}(thf)$_2$] (2.06(9)Å) and in [Li$_2${(*R*)-PEPIA] (1.933(7)-2.142(7)Å). The N-P-N bond angle of the ligand (*R*)-HPE^{Dipp}PIA (116.7(2)°) significantly decreases upon coordination with lithium centres as previously observed in complex **4** (N1-P1-N2 105.6(2)° and N3-P2-N4 104.7(2)°).

In order to obtain the sodium analogue [{(*R*)-PE^{Dipp}PIA}$_2$Na$_2$] (**5**), similar base elimination reaction between (*R*)-HPE^{Dipp}PIA and NaN(SiMe$_3$)$_2$ was carried out in an equimolar ratio (Scheme 3.1.5). The M···C interaction is evidenced in the ^{1}H NMR spectrum by resonances shifted upfield to the aromatic region δ= 7.67-6.56 in **5** *vs.* δ= 7.78-6.92 ppm in (*R*)-HPE^{Dipp}PIA. A doublet of quartets corresponding to the Ph(C*H*)CH$_3$ protons is observed at δ = 4.01 ppm with coupling constants of $^3J_{PH}$ = 25.18 Hz and $^3J_{HH}$ = 6.35 Hz, respectively. The Ph(CH)C*H*$_3$ protons appear as a doublet at δ= 1.44 ppm ($^3J_{HH}$ = 6.38 Hz). Due to the hindered rotation of the Dipp groups, the corresponding methine and methyl protons (*i.e.* C*H*(CH$_3$)$_2$ and CH(C*H*$_3$)$_2$) appear as broad and multiplet signals at δ = 3.23 ppm ($\Delta\nu_{1/2} \approx$ 186.3 Hz) and δ = 0.99-0.27 ppm, respectively.

Complex **5** crystallizes in the chiral orthorhombic space group $P2_12_12_1$ with two molecules in the asymmetric unit (Figure 3.1.9). Complex **5** forms a fused tricyclic ladder motif, similar to that observed in **4**.

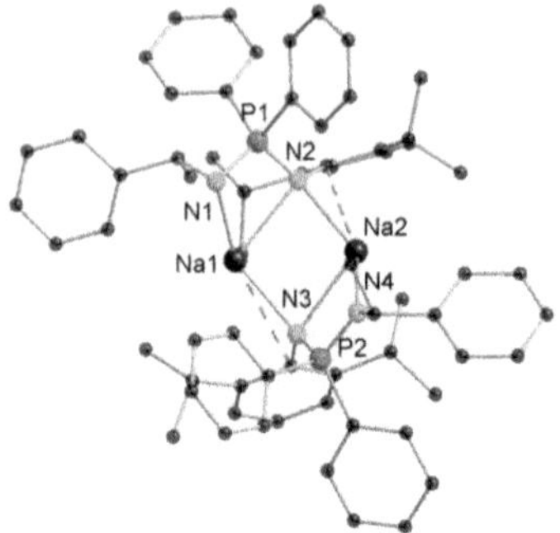

Figure 3.1.9 Molecular structure of **5** in the solid state. Hydrogen atoms are omitted for clarity. Selected bond lengths [Å] and angles [°]: Na1-N1 2.313(4), Na1-N2 2.516(4), Na1-N3 2.390(4), Na2-N2 2.379(4), Na2-N3 2.519(4), Na2-N4 2.292(4), P1-N1 1.591(4), P1-N2 1.601(4), P2-N3 1.606(4), P2-N4 1.585(4); N1-Na1-N2 63.93(13), N1-Na1-N3 135.1(2), N2-Na1-N3 100.56(14), N2-Na2-N3 100.80(14), N2-Na2-N4 136.7(2), N3-Na2-N4 64.44(13), Na1-N2-Na2 77.78(12), Na1-N3-Na2 77.52(12), N1-P1-N2 106.8(2), N3-P2-N4 107.4(2). Torsion angle [°]: N1-P1-P2-N4 57.35(2). M-M and P-P distance [Å]: Na1-Na2 3.075(2), P1-P2 5.59(2).

The dimeric form of **5** is different from the ion pair observed in case of $[\{Ph_2P(NSiMe_3)_2\}_2Na][Na(thf)_6]$.[32] The sodium atoms are both coordinated in chelating as well as bridging bonding modes by (*R*)-HPE^{Dipp}PIA. To minimize the steric repulsion, the two Dipp groups are arranged in *trans* fashion. In contrast to complex **4**, where no M···arene interaction could be observed, the sodium atoms in complex **5** show an η^1 interaction with the *ipso* carbon (*i.e.* adjacent to nitrogen) of the Dipp groups. This interaction might be due to larger ionic radius of sodium compared to lithium. The Na···arene interaction observed in case of $[Na_2\{(R)\text{-PEPIA}\}_2]$,[103b] and the achiral analogue, $[Na\{(2,6\text{-}iPr_2C_6H_3N)_2P(Ph_2)\}]_2$,[29] is higher as compared to that in complex **5**. The NPN angle of the ligand (*R*)-HPE^{Dipp}PIA, (116.7 (2)°) significantly decreases upon coordination to the sodium metals in complex **5** (N1-P1-N2 106.8(2)°, N3-P2-N4 107.4(2)°).

3.1.3.3 Synthesis of $[\{(R)\text{-NEPIA}\}Li(thf)_2]$ (6) and $[\{(R)\text{-NEPIA}\}_2M_2]$ (M = Na (7), K(8))

Further, the study of the coordination pattern of chiral iminophosphonamides with alkali metals was extended to the new ligand (*R*)-HNEPIA. In this regard, deprotonation of (*R*)-HNEPIA with an equimolar amount of $LiN(SiMe_3)_2$ led to the formation of monomeric lithium complex [{(*R*)-NEPIA}Li(thf)$_2$] (**6**) (Scheme 3.1.6). The monomeric form of **6** is in contrast to the above discussed dimeric forms of [{(*R*)-PEPIA}$_2$Li$_2$][103b] and [{(*R*)-PE^{Dipp}PIA}$_2$Li$_2$] (**4**). Complex **6** is soluble in THF, toluene and diethyl ether whereas insoluble in nonpolar solvents such as *n*-pentane.

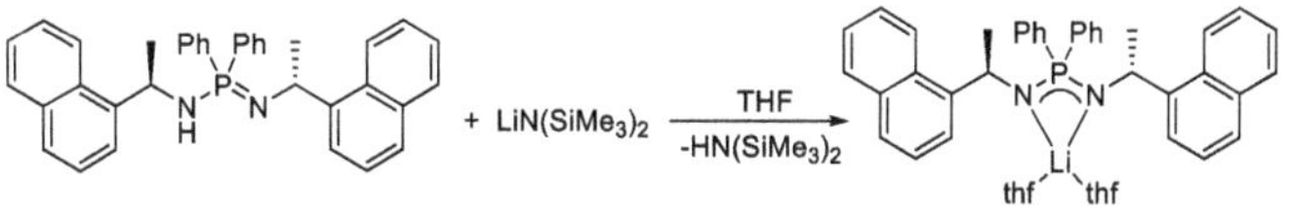

Scheme 3.1.6 Synthesis of [{(*R*)-NEPIA}Li(thf)$_2$] (**6**).

In the ^{1}H NMR spectrum of complex **6**, a characteristic resonance for naph(C*H*)CH$_3$ could be observed as a doublet of quartets at δ = 5.17 ppm ($^3J_{PH}$ = 19.79 Hz and $^3J_{HH}$ = 6.2 Hz, respectively). A doublet at δ = 1.17 ppm ($^3J_{HH}$ = 6.32 Hz) is attributed to naph(CH)C*H*$_3$. The identity of complex **6** was further confirmed by analysis of the ^{7}Li{^{1}H} NMR spectrum, which shows a single resonance at δ = 2.7 ppm. The ^{31}P{^{1}H} NMR spectrum of complex **6** exhibits a single resonance at δ = 30.4 ppm, significantly shifted downfield as compared to that of the starting precursor (*R*)-HNEPIA (δ = 3.6 ppm). The complete deprotonation of (*R*)-HNEPIA is also indicated by disappearance of the N-H stretching band in IR spectrum of complex **6**.

Single crystals suitable for X-ray analysis were obtained from a saturated THF solution of **6**. Interestingly, attempts to grow crystals in noncoordinating solvents such as toluene or hot heptane did not result in successful crystallisation, suggesting that coordination of THF is desirable to isolate **6** in a crystalline state. Complex **6** crystallizes in the chiral monoclinic space group *P*2$_1$ with one molecule in the asymmetric unit (Figure 3.1.10). The lithium centre in complex **6** adopts a distorted tetrahedral geometry and the coordination sphere is taken up by two nitrogens of {(*R*)-NEPIA}$^-$ ligand and two thf molecules. In contrast to the dimeric lithium complex [{(*R*)-PEPIA}$_2$Li$_2$], complex **6** resembles the previously known lithium complex, [{Ph$_2$P(Me$_3$SiN)$_2$}Li(thf)$_2$],[32] where the four-membered N$_2$PLi metallacycle exists in a kite shape. The lithium centre in complex **6** does not lie in the NPN plane as shown by a slight deviation of the dihedral angle of 9.086(6)°. This deviation might be caused by the repulsion between the solvent molecules (thf) and the naphthyl rings. The P-N bond lengths of 1.592(2) Å and 1.596(2) Å and Li-N bond lengths of 2.020(4) and 2.027(5) Å indicate that the lithium centre is located exactly in the bisection of the N-P-N backbone.

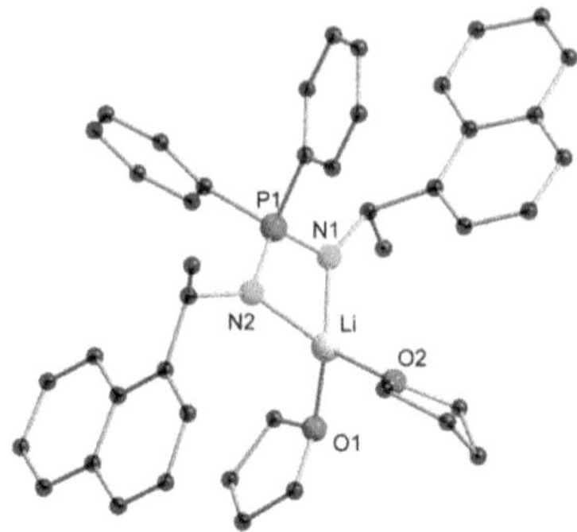

Figure 3.1.10 Molecular structure of **6** in the solid state. Hydrogen atoms are omitted for clarity. Selected bond lengths [Å] and angles [°]: Li-N1 2.020(4), Li-N2 2.027(5), Li-O1 1.986(5), Li-O2 1.965(5), 1.986(5), P1-N1 1.592(2), P1-N2 1.596(2); N1-Li-N2 76.9(2), O1-Li-O2 100.2(2), N1-Li-O2 115.1(2), O1-Li-N2 114.5(2), O1-Li-N1 126.0(3), O2-Li-N2 125.6(3), N1-P1-N2 104.20(10) Torsion angle [°]: P1-N1-Li-O2 105.33(1), P1-N2-Li-O1 117.23(1).

The Li-O bond lengths (1.986(5) and 1.965(5) Å) and the N1-Li-N2 angle (76.9(2)°) are comparable to those reported in literature.[25c]

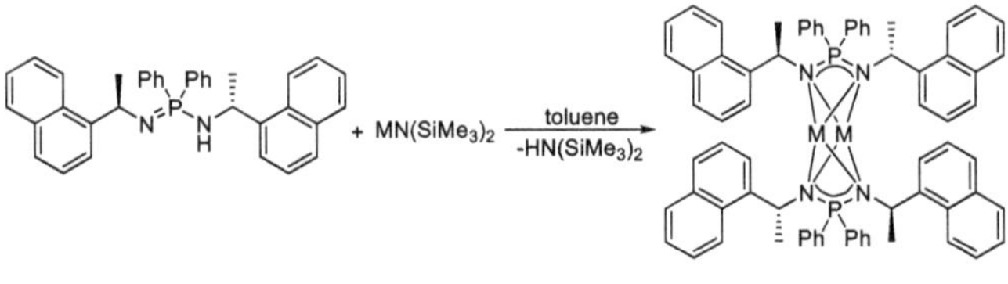

+ $MN(SiMe_3)_2$ toluene $-HN(SiMe_3)_2$

M = Na (**7**), K (**8**)

Scheme 3.1.7 Synthesis of [{(*R*)-NEPIA}$_2$M$_2$] (M = Na(**7**), K(**8**)).

(*R*)-HNEPIA was further treated with and equimolar amount of $MN(SiMe_3)_2$ (M = Na, K), which led to the facile formation of [{(*R*)-NEPIA}$_2$M$_2$] (M = Na (**7**), K (**8**)) (Scheme 3.1.7).

The appearance of a single set of signals in the ^{1}H NMR spectrum for both **7** and **8** suggests a symmetric coordination of {(*R*)-NEPIA}$^-$ with the metal centres in solution. In the ^{31}P{^{1}H} NMR spectrum of complexes **7** and **8** show a single peak at δ = 24.8 ppm and δ = 18.5 ppm, respectively, which is shifted upfield compared to **6** (δ = 30.4 ppm).

The identities of complex **7** and **8** were further established by single crystal X-ray diffraction studies. Both complexes are isomorphous and crystallize in the chiral monoclinic space group $P2_1$ with one molecule in the asymmetric unit (Figure 3.1.11).

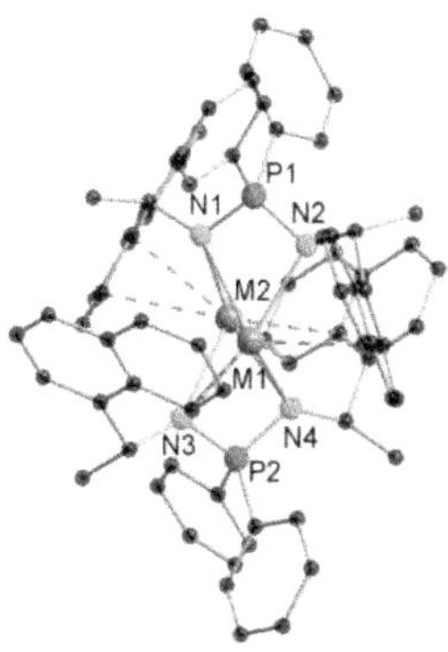

Figure 3.1.11 Molecular structures of **7** (M = Na), and **8** (M = K) in the solid state. All of the hydrogen atoms are omitted for clarity. Selected bond lengths [Å] and angles [°]: For **7**: Na1-N1 2.753(2), Na1-N2 2.359(2), Na1-N3 2.593(2), Na1-N4 2.513(2), Na2-N1 2.400(2), Na2-N3 2.440(2), Na2-N4 2.826(2), P1-N1 1.601(2), P1-N2 1.590(2), P2-N3 1.602(2), P2-N4 1.603(2); N1-Na1-N2 59.34(7), N4-Na1-N3 59.49(7), N4-Na1-N1 95.62(7), N2-Na1-N3 104.98(7), N2-Na1-N4 147.99(8), N1-Na2-N3 122.10(8), N1-Na2-N4 96.35(7), N3-Na2-N4 56.96(7), Na2-N1-Na1 64.70(6), Na2-N3-Na1 66.81(6), Na1-N4-Na2 62.27(6), N2-P1-N1 106.32(10), N3-P2-N4 104.48(10). Torsion angle [°]: N1-P1-P2-N3 136.56(2). M-M and P-P distance [Å]: Na1-Na2 2.77(2), P1-P2 5.72 (14). M···C coordination interactions [Å]: Na1···C4 3.102, Na1···C45 2.782, Na2···C68 3.056, Na2···C69 3.012, Na2···C27 2.788, Na2···C28 2.854. For **8**: K1-N2 2.831(4), K2-N2 2.996(3), K1-N3 3.198(3), K1-N4 2.762(4), K2-N1 2.783(3), K2-N3 2.790(3), K2-N4 2.927(4), P1-N1 1.592(4), P1-N2 1.607(3), P2-N3 1.588(4), P2-N4 1.597(3); N2-K1-N3 94.50(9), N4-K1-N2 116.19(10), N4-K1-N3 50.03(9), N2-K1-N3 94.50(9), N4-K1-N2 116.19(10), N1-K2-N3 140.22(10), N1-K2-N4 102.53(10), N3-K2-N4 52.92(10), K2-N3-K1 64.81(7), N1-K2-N4 102.53(10), N1-P1-N2 108.1(2), N3-P2-N4 106.4(2). Torsion angle [°]: N1-P1-P2-N3 146.45(2). M-M and P-P distance [Å]: K1-K2 3.23(2), P1-P2 6.25 (2). M···C coordination interactions [Å]: K1···C39 3.102, K1···C44 2.999, K2···C27 3.23, K2···C28 3.18, K2···C68 3.24, K2···C69 2.96, K2···C70 3.38.

The solid-state structures of **7** and **8** indicate corresponding dimeric arrangements in which the central metals show both bridging and chelating bonding modes for the [(*R*)-NEPIA]$^-$ ligand, which is in consistence to the above-mentioned dimeric complexes (**1-5**). Complex **7** exhibits (η^1, η^1) and (η^2, η^2) coordination modes for Na1 and Na2, respectively, whereas complex **8** shows (η^3, η^3) and (η^3, η^2) bonding modes of the naphthyl rings attached to the chiral carbon with K1 and K2, respectively. The M···C interactions for complex **7** are in the range 2.783(3)-3.056 Å whereas, for complex **8**, they are in the range of 2.962(2)-3.437 Å, both of which are well comparable with the reported values.[108,111] The average M-N bond length of 2.87(2) Å in **8** (M = K) is longer than that in **7** (2.55(2) Å, M = Na), which is likely due to a larger ionic radius of potassium compared to sodium. The increased ionic radius of the central metal also affects the N1-P1-P2-N3 torsion angles, with an increase from 136.56(2)° (**7**) to 146.45(2)° (**8**).

3.1.4 Photoluminescence Study

The photoluminescence (PL) studies of the alkali metal complexes [{(*R*)-PEPIA}$_2$M$_2$] (M = Li, Na, K (**1**), Rb (**2**) and Cs (**3**)) were carried out by Dr. Thomas Feuerstein and Dr. Sergei Lebedkin (group of Prof. Manfred Kappes).[103b]

Interestingly, when the ligand (*R*)-HPEPIA and the corresponding alkali metal complexes were exposed under a 365 nm UV lamp at ambient conditions, blue green photoluminescence (PL) was observed (left, Figure 3.1.12).

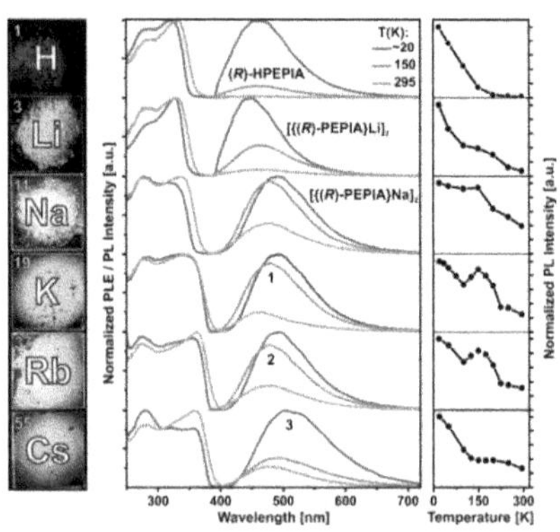

Figure 3.1.12 Left: photographs of solid (polycrystalline) compounds (*R*)-HPEPIA, [{(*R*)-PEPIA}$_2$Li$_2$], [{(*R*)-PEPIA}$_2$Na$_2$] and **1-3** under UV excitation at ambient temperature. Middle: photoluminescence excitation (PLE) and emission (PL) spectra of the compounds at 20, 150 and 295 K. Right: integrated PL intensities plotted against the temperature in the range of 20 to 295 K.

A comparison of the PL behaviour at different temperatures was carried out. In this regard, the PL spectra of the complexes were recorded at 20, 150 and 295 K, respectively. The PL emission and PL excitation (PLE) spectra as well as the temperature dependencies of the integrated PL intensity for the different metal complexes are shown in Figure 3.1.12. The emission spectra of the compounds are broad with a width of about 100-120 nm (FWHM). The PL intensity of each compound increases upon lowering the temperature from 295 to 20 K (middle, Figure 3.1.12). At 295 K, broad emission bands for the ligands (*R*)-HPEPIA, [{(*R*)-PEPIA}$_2$Na$_2$] and the complexes **1-3** are centred at 450, and 462-480 nm. Upon lowering the temperature to 20 K, the emission bands show moderate red shifts to 462 and 490-505 nm, respectively. In contrast, the Li complex [{(*R*)-PEPIA}$_2$Li$_2$] shows a small blue-shift from 462 to 450 nm.

At 20 K, the PL of (*R*)-HPEPIA is bright but decreases exponentially with rising temperature. Finally, at 295 K, it decreases by a factor of 100 and a quantum yield of ϕ_{PL} (295) = 0.37% has been determined (Table 3.1.2). The fast decay in the 2-3 ns range suggests that the PL of (*R*)-HPEPIA is fluorescence. Interestingly, the alkali metal complexes of (*R*)-HPEPIA show a similar PL behaviour to that observed for the (*R*)-HPEPIA ligand itself. The alkali metal complexes however show quite distinct temperature-dependent PL intensities. Interestingly, [{(*R*)-PEPIA}$_2$Na$_2$], [{(*R*)-PEPIA}$_2$K$_2$] (**1**) and [{(*R*)-PEPIA}$_2$Rb$_2$] (**2**) show non-monotonic decay of the PL intensity in the temperature range from 20 to 295 K. Especially in the temperature region of 120-200 K, an increase in the PL intensity could be detected. In contrast, in this temperature region, the complexes [{(*R*)-PEPIA}$_2$Li$_2$] and [{(*R*)-PEPIA}$_2$Cs$_2$] (**3**) show almost monotonic decay of the PL. At room temperature, the observed PL quantum yield of the dimeric complexes [{(*R*)-PEPIA}$_2$M$_2$] (M = Li, Na, K (**1**), Rb (**2**) and Cs (**3**)) are 8, 36, 21, 21 and 3 % which increase to 80, 91, 92, 64 and 13 % at 20 K, respectively.

Table 3.1.2: Characteristic spectroscopic parameters for the photoluminescence (PL) of compounds.

Compound	**(*R*)-HPEPIA**	**[{(*R*)-PEPIA}$_2$Li$_2$]**	**[{(*R*)-PEPIA}$_2$Na$_2$]**	**1**	**2**	**3**
Central metal	-	Li	Na	K	Rb	Cs
λ_{PLE}[a] [nm]	320	330	330	350	365	350
λ_{PL}(20 K)[b] [nm]	462	465	490	490	490	505
λ_{PL}(295 K) [nm]	450	450	476	462	476	480
FWHM(20 K)[c] [nm]	106	91	111	105	103	131
FWHM(295 K) [nm]	114	118	109	90	110	128
$\lambda_{PLE}(\phi)$[d] [nm]	330	330	330	350	350	350
ϕ(RT)[e] [%]	0.37	8	36	21	21	3
τ_1(20 K)[f] [ns]	2-3	<10	<10	<10	<10	<10
τ_2(20 K, Phosph.) [ms]	-	6.0	4.3	8.1	9.0	0.6/5.1
τ_2(295 K, DF) [µs]	-	4.1	6.9	12.5	14.8	1.6/9
$\Delta E(S_1-T_1)$ [meV]	-	-	79	73	90	-

[a]: Excitation wavelength used for the recording of the PL and PLE spectra at 20 and 295 K; [b]: Band maxima of the PL spectra; [c]: FWHM = Full Width at Half Maximum of the PL; [d]: excitation wavelength used for the determination of the quantum yields; [e]: PL quantum yields at room temperature, determined according to ref.[112]; [f]: Lifetime of the PL, determined by monoexponential curve fitting to the measured decay curves, Phoshp = phosphorescence, DF = delayed fluorescence.

It should be noted that the lower quantum yield of the caesium analogue (**3**), compared to the other complexes, might be attributed to the cocrystallization of dimeric (**3$_d$**) and polymeric (**3$_p$**) units. The PL decay of the alkali metal complexes show a major long-lived component along with an almost negligible fast decay component (<1%), as seen for (*R*)-HPEPIA. The long-lived component of the PL decays monoexponentially and varies with the temperature. For each metal complex, the decay time varies from a few ms below 100 K to a few μs above 200 K. Especially the potassium complex **1** shows a PL decay with τ = 8.14 ms at 20 K and a decay time of τ = 14.8 μs at 295 K. The millisecond long-lived component of the decay can be assigned to phosphorescence from the T_1 triplet state to the S_0 ground state. A slightly faster long-lived component of the order of μs at elevated temperature is apparently different and is assigned to thermally activated delayed fluorescence (TADF). The TADF occurs due to close proximity in energy of the T_1 and the S_1 state. Due to this small energy gap, the T_1 state populates S_1 *via* reversible intersystem crossing (rISC).[113] This delayed fluorescence of [{(*R*)-PEPIA}$_2$M$_2$] (M = Na, K (**1**) and Rb (**2**)) has been investigated in detail and illustrated for Na analogue (Figure 3.1.13).

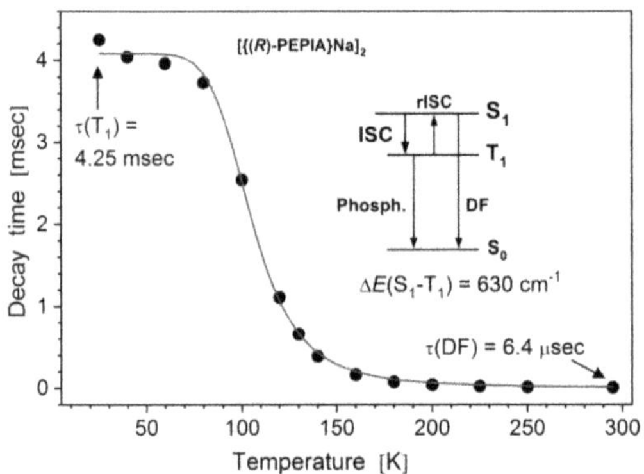

Figure 3.1.13 PL decay time of solid (polycrystalline) Na complex [{(*R*)-PEPIA}$_2$Na$_2$] *vs* temperature. The (monoexponential) decay curves were recorded at 490 nm using a ns-pulsed laser for excitation at 337 nm. The insert scheme depicts a delayed fluorescence (DF) process due to thermally activated reverse intersystem crossing (rISC) from T_1 to S_1 state. At low temperatures, the emission is phosphorescence from the T_1 state. The blue line shows the fit according to equation (1), yielding $\tau(T_1)$ = 4.06 ms, $\tau(S_1)$ = 230 ns and S_1-T_1 energy separation of 630 cm^{-1} K.

Below 100 K, the sodium complex shows phosphorescence (with $\tau(T_1)$ = 4.06 ms at 40 K), whereas, upon increasing the temperature to above 150 K, it turns into delayed S_1 fluorescence (with the effective time τ = 6.4 μs at 295 K). In this region, the S_1 state is repopulated by reverse intersystem crossing (rISC) competing with intersystem (ISC) transitions. This observation correlates with the non-monotonic behaviour of integral PL intensity for [{(*R*)-PEPIA}$_2$Na$_2$], **1** and **2**. For TADF, the energy separation ΔE between S_1 and

T_1 states can be estimated from the (monoexponential) decay times $\tau(T)$ using the simple model of thermally equilibrated states, according to equation 1.[114]

$$\tau(T) = \frac{3 + \exp(-\frac{\Delta E}{kT})}{\frac{3}{\tau(T1)} + \frac{1}{\tau(S1)}\exp(-\frac{\Delta E}{kT})} \qquad \text{(equation 1)}$$

where kT is the thermal energy, $\tau(S_1)$ and $\tau(T_1)$ are the intrinsic fluorescence and phosphorescence lifetimes. A factor of 3 takes into account that there are three T_1 substates. It should be noted that this model does not account for temperature-dependent relaxation (both radiative and non-radiative) and thus is expected to be more accurate for TADF compounds with a moderate temperature dependence of the PL intensity. Based on the equation 1 and the experimental data, the calculated $\Delta E(S_1\text{-}T_1)$ yielded as 630, 590, 730 cm^{-1} which is 78, 73 and 90 meV, for the Na, K (**1**) and Rb (**2**) complexes, respectively. These values of $\Delta E(S_1\text{-}T_1)$ are in the expected range for molecules to show TADF. These values of $\Delta E(S_1\text{-}T_1)$ are also in agreement with the values obtained from DFT calculations.

As already mentioned, the ligand (*R*)-HPEPIA mostly shows fluorescence due to relatively inefficient depopulation of T_1 state via rISC. In contrast, the dimeric alkali metal complexes exhibit both phosphorescence and TADF, although with varying PL efficiencies. Further, the heavy metal effect for efficient ISC and rISC can be ruled out for the lighter alkali metals such as Li, Na or K. The interesting PL of these metal complexes are attributed to their dimeric structure. For instance, high triplet yields have been achieved in orthogonal dimers of Bodipy fluorescent dye molecules, which normally produce no triplets.[115] These dimeric complexes represent the first example of TADF emitters based on iminophosphonamides as well as of dimeric TADF molecules to the best of our knowledge.

Further, in order to gain more insight into the phenomenon, a quantum chemical calculation was performed by Dr. Ralf Köppe.

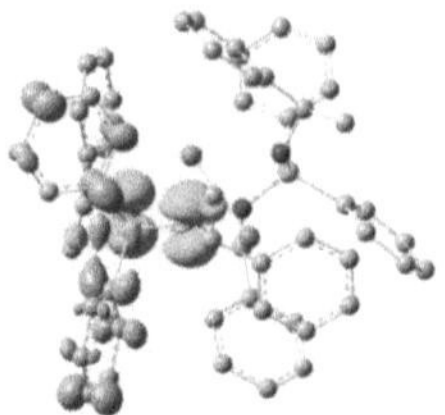

Figure 3.1.14 Non-relaxed difference electron density plot under excitation of the compound [{(*R*)-PEPIA}$_2$Na$_2$] (isosurface value ± 0.004, the light blue and dark violet areas correspond to the regions of decreased and increased electron density upon excitation, respectively). The Na atoms are plotted in violet.

Based on the calculations, it can be concluded that the electron density present on the metal cation practically does not involve into the excitation and emission transition. The calculations further suggest that the highest occupied molecular orbital (HOMO) in (*R*)-HPEPIA and in the corresponding alkali metal complexes is composed of p-type electron lone pairs on the N atoms, whereas the lowest unoccupied molecular orbital (LUMO) is of antibonding π^*-character and mostly contributed by phenyl groups connected to the phosphorus atoms. As expected, the local excitation in nearly degenerate energy levels is found to be either in the monomeric unit or in a linear combination of a monomeric and dimeric unit. This is further illustrated by the frontier orbitals (HOMO, HOMO-1, LUMO and LUMO+1) (Figure 3.1.15).

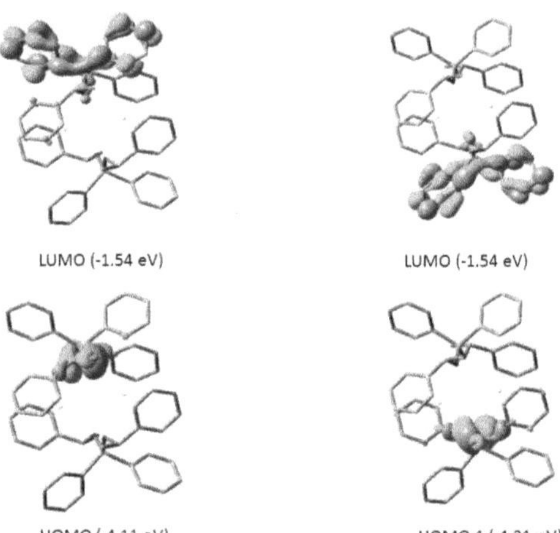

Figure 3.1.15 Plots of frontier MOs of **2** (MO energy values obtained from the S_0 ground state structure). In the difference density plots, the light blue and dark violet areas correspond to the regions of decreased and increased electron density upon excitation, respectively (isosurface values ±0.04 (HOMO, LUMO), ±0.002 (difference density plot)).

The calculated values of the energy separations lie in the range of 270-740 cm^{-1}, which is in good agreement with the experimental estimates for the Na, K and Rb complexes of (*R*)-HPEPIA.

3.2 Alkaline earth metal and ytterbium (II) complexes of chiral iminophosphonamines

3.2.1 Introduction

Owing to their environmentally friendly nature and low toxicity, alkaline earth metals are widely employed as catalysts in various organic reactions.[116] Most of these chemical transformations include hydrophosphination of alkenes and alkynes,[9,117] coupling of alkynes with carbodiimides,[118] Tishchenko reactions,[119] ring opening polymerizations,[120] hydroborations,[92d,121] hydroaminations,[94,121c,122] and cross-dehydro-couplings.[123] Until the 1970s, most of the effort to synthesize alkaline earth metal complexes involved functionalization of organomagnesium complexes.[124] Since then, many other tailor-made ligands have been utilized to control metal nuclearity, coordination number, geometry and reactivity of the respective complexes.[125] However, the literature known alkaline earth metal complexes based on achiral iminophosphonamides are limited to only a few ligand backbones as discussed in section 1.2.2. [33-34,36] Albeit, alkaline earth metal complexes based on chiral iminophosphonamides (having chiral substituents only at one nitrogen centre) are only known in the case of a heteroleptic calcium complex.[101b] Therefore, in this work, alkaline earth metal (Mg, Ca) complexes of chiral iminophosphonamines (Chart 3.2.1) are synthesized. Further, the catalysis and photoluminescence behaviour of these complexes has been studied.

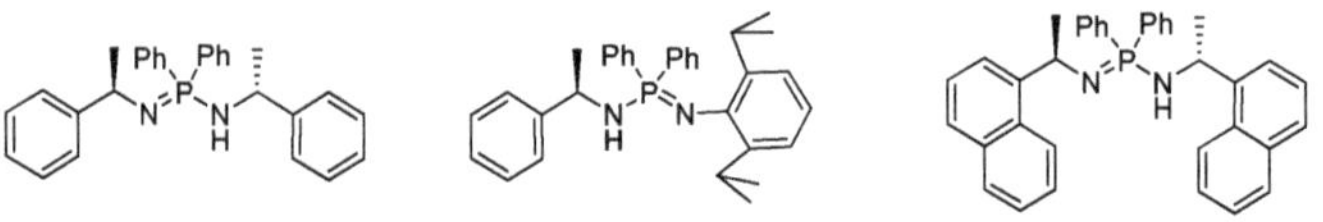

(*R*)-HPEPIA (*R*)-HPE^{Dipp}PIA (*R*)-HNEPIA

Chart 3.2.1 Utilised enantiopure iminophosphonamine ligands to synthesize group 2 complexes.

3.2.1.1 Synthesis of enantiopure alkaline earth metal and divalent ytterbium complexes (9-14)

In the course of this study, the iminophosphonamide complex of Ca(II), [{(*R*)-PEPIA}$_2$Ca] (**9**), was prepared by reacting the pre-synthesized [Ca{(NSiMe$_3$)$_2$}$_2$(thf)$_2$],[126] with the iminophosphonamine (*R*)-HPEPIA (Scheme 3.2.1). It is noteworthy, that, irrespective of the (*R*)-HPEPIA and [Ca{(NSiMe$_3$)$_2$}$_2$(thf)$_2$] molar ratio, the product formed is always the homoleptic complex [{(*R*)-PEPIA}$_2$Ca] and the yield can be improved by using a ligand to metal molar ratio of 2:1.

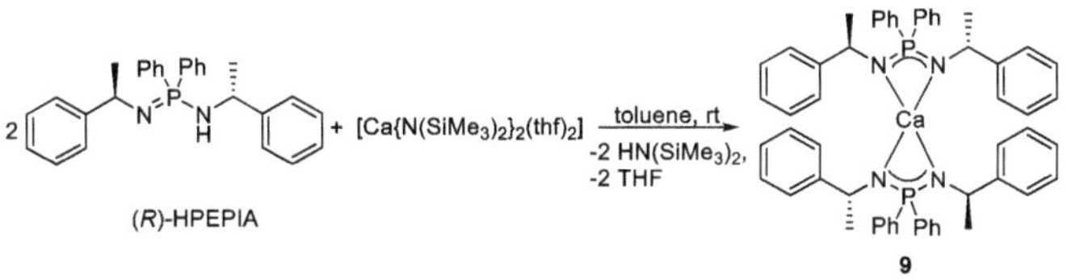

2 (*R*)-HPEPIA + [Ca{N(SiMe$_3$)$_2$}$_2$(thf)$_2$] —(toluene, rt; -2 HN(SiMe$_3$)$_2$, -2 THF)→ **9**

Scheme 3.2.1 Synthesis of complex [{(*R*)-PEPIA}$_2$Ca] (**9**).

This suggests that the (*R*)-PEPIA ligand is not sufficiently bulky to stabilise the heteroleptic [{(*R*)-PEPIA}CaN(SiMe$_3$)$_2$] product. The homoleptic calcium complex **9** was characterized by multinuclear NMR spectroscopy, IR and elemental analysis. All of the analytics hint towards the formation of **9**. The ^{1}H NMR spectrum of complex **9** shows a single set of resonances, suggesting its symmetric nature in solution (Figure 3.2.1). The characteristic signal for Ph(C*H*)CH$_3$ moiety of complex **9** appear as a doublet ($^{3}J_{PH}$ = 19.58 Hz) of quartet ($^{3}J_{HH}$ = 6.49 Hz) at δ = 4.04 ppm, and corresponding methyl protons Ph(CH)C*H*$_3$ appear as a doublet at δ = 1.38 ppm with a coupling constant of $^{3}J_{HH}$ = 6.49 Hz.

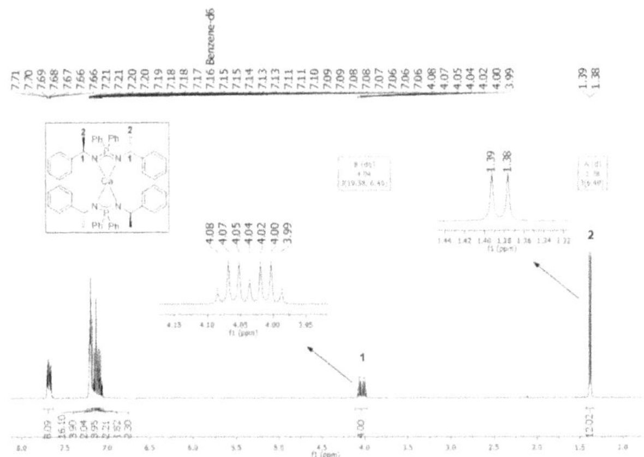

Figure 3.2.1 ^{1}H NMR spectrum (400 MHz, C_6D_6, 295K) of (**9**).

However, the methine (Ph(C*H*)CH$_3$) and methyl (Ph(CH)C*H*$_3$) protons of the ligand (*R*)-HPEPIA appear to show two sets of signals at δ = 4.73, 4.38 ppm and δ = 1.66, 1.13 ppm, respectively.

The purity of complex **9** was further confirmed by appearance of a single resonance at δ = 24.5 ppm in $^{31}P\{^{1}H\}$ NMR spectrum, which is significantly shifted downfield compared to δ = 2.7 ppm observed for (*R*)-HPEPIA. In the $^{13}C\{^{1}H\}$ NMR spectrum, the resonances at δ = 54.1 ppm and δ = 28.4 ppm are assigned to the methine (Ph(*C*H)CH$_3$) and methyl (Ph(CH)*C*H$_3$) groups, respectively. Further, disappearance of N-H stretch in the IR spectrum of complex **9** suggest complete deprotonation of the ligand (*R*)-HPEPIA.

The molecular structure of complex **9** was determined by single crystal X-ray diffraction. Complex **9** crystallizes in the chiral orthorhombic space group *C*2 with two halves of the molecule in the asymmetric unit. The coordination polyhedron of the central calcium metal can be described as a distorted tetrahedron (Figure 3.2.2).

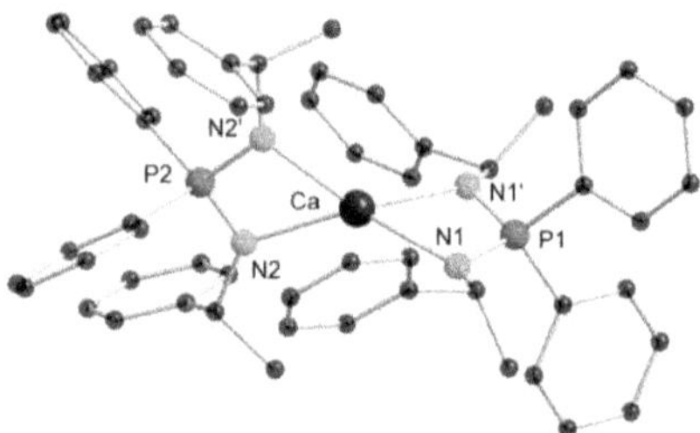

Figure 3.2.2 Molecular structure of complex **9** in the solid state. All hydrogen atoms are omitted for clarity. Selected bond lengths (Å) and bond angles [°]: Ca-N1 2.365(3), Ca-N2 2.344(2), Ca-P1 2.9884(14), Ca-P2 2.9801(14), P1-N1 1.606(3), P2-N2 1.605(2); P1-Ca-P2 180.0, N1-Ca-N2 137.17(9), N1'-Ca-N2' 137.18(9) N2-Ca-N2' 64.76(11), N1-Ca-N1' 64.64(13), N2-Ca-N1' 133.94(9), N1-Ca-N2' 133.94(9), N1-P1-N1' 103.9(2), N2-P2-N2' 102.9(2).

This deviation of complex **9** from ideal tetrahedral geometry is attributed to the chelating effect of the ligand (*R*)-HPEPIA, which offers bite angles of N2-Ca-N1 137.17(9)° and N2'-Ca-N1' 137.18(9)°. The Ca-N bond lengths (Ca-N1 2.365(3) Å and Ca-N2 2.344(2) Å) are in a comparable range and also in consistency with literature reports of similar compounds.[127] The identical P-N bonds (P1-N1 1.606(3) Å, P2-N2 1.605(2) Å) observed in complex **9** indicate charge delocalization in the PN$_2$ fragment. Furthermore, the calcium atom lies exactly in the plane of the two NPN backbones with an exactly linear P1-Ca-P2 vector (180°), and the orthogonal of the two N$_2$P planes intersect at an angle of 86.04(3)°. This is in contrast to the bent P1-Ca-P2 vector (144.3°) observed in case of [(thf)Ca{(NSiMe$_3$)$_2$PPh$_2$}$_2$], realizing a distorted trigonal bipyramidal geometry.[33] Upon complex formation, the chelating angle of the (*R*)-HPEPIA ligand (121.6(2)°) significantly decreases to (N1-P1-N1′ 103.9(2)° and N2-P2-N2′ 102.9(2)°) in **9**.

Further, the unsymmetrical ligand (*R*)-HPE^{Dipp}PIA was reacted with [Mg{N(SiMe$_3$)$_2$}$_2$(thf)$_2$] in a 1:1 molar ratio, which resulted in the expected heteroleptic magnesium complex [{(*R*)-PE^{Dipp}PIA}{N(SiMe$_3$)$_2$}Mg(thf)] (**10**) (Scheme 3.2.2). Favorable formation of **10** is most likely due to the bulkier nature of the ligand (*R*)-HPE^{Dipp}PIA. The ^{1}H NMR spectrum of **10** is consistent with the mono-solvated thf adduct and shows a characteristic doublet for Ph(CH)CH_3 at δ = 1.46 ppm with a coupling constant of $^3J_{HH}$ = 6.7 Hz. A characteristic singlet for the methyl protons of N(SiMe_3)$_2$ is detected at δ = 0.37 ppm.

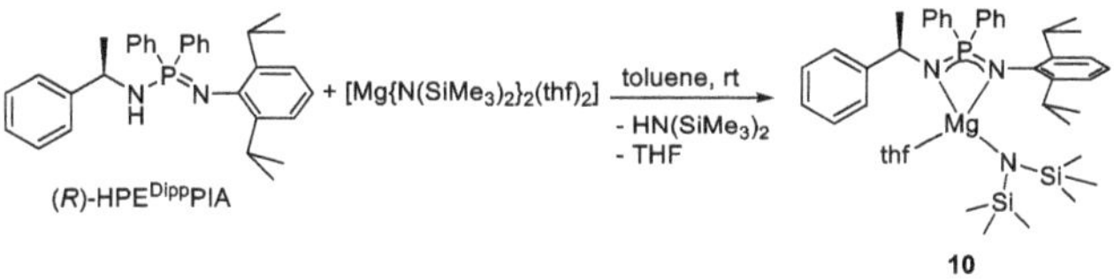

Scheme 3.2.2 Synthesis of complex [{(*R*)-PE^{Dipp}PIA}{N(SiMe$_3$)$_2$}Mg(thf)] (**10**).

Further, the complete deprotonation of (*R*)-HPE^{Dipp}PIA was confirmed by disappearance of N-H stretching frequency in the IR spectrum of complex **10**.

Single crystals suitable for X-ray analysis of **10** were grown from a saturated solution in *n*-pentane. Complex **10** crystallizes in the chiral orthorhombic space group $P2_12_12_1$ with one molecule in the asymmetric unit (Figure 3.2.3). The geometry around the Mg centre in **10** is a distorted tetrahedral where the four coordination sites of the magnesium centre are occupied by three nitrogen and one oxygen atom. The P-N bond lengths of complex **10** are in the expected range and within estimated standard deviations, suggesting complete charge delocalization in the PN$_2$ fragment. Further, the Mg-N bond lengths of complex **10** are in the range between 2.005(2) Å and 2.126(2) Å, which are in the similar range (2.06(2)-2.10(1) Å) of related magnesium complexes.[13a,33] The central magnesium is almost co-planar with the NPN backbone, having a slight deviation with a dihedral angle of 1.68(2)°.

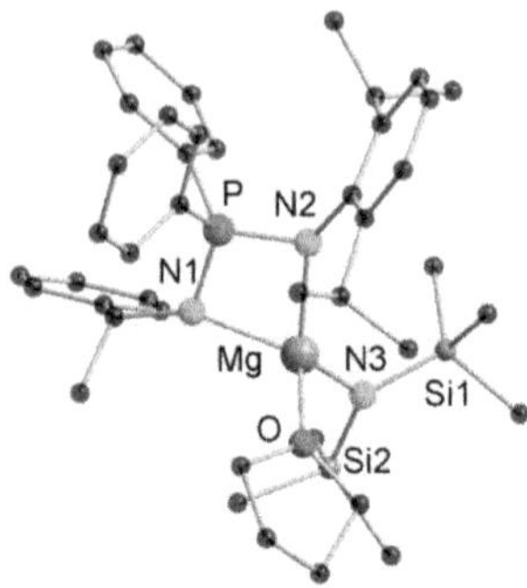

Figure 3.2.3 Molecular structure of complex **10** in the solid state. All hydrogen atoms are omitted for clarity. Selected bond lengths (Å) and bond angles [°]: Mg-N1 2.080(2), Mg-N2 2.126(2), Mg-N3 2.005(2), Mg-O 2.05(2), P-Mg 2.7110(10), P-N1 1.614(2), P-N2 1.614(2), Si1-N3 1.705(2), Si2-N3 1.704(2); N1-Mg-N2 72.98(7), O-Mg-N1 105.24(8), O-Mg-N2 116.32(8), N1-Mg-N3 133.66(8), N2-Mg-N3 125.01(9), N1-P-N2 101.57(10), Si1-N3-Si2 123.06(12).

This slight inclination of the magnesium center most likely minimizes the steric repulsion between Dipp and $N(SiMe_3)_2$ group. Upon complex formation with magnesium, the chelating angle of (*R*)-HPE^{Dipp}PIA (116.7(2)°) significantly decreases to 101.57(10)° in **10**.

Moreover, performing a base elimination reaction between (*R*)-HPE^{Dipp}PIA and $[M\{N(SiMe_3)_2\}_2(thf)_2]$ (M = Mg, Ca) in 2:1 ratio afforded the homoleptic complexes, $[\{(R)$-PE^{Dipp}PIA$\}_2$Mg] (**11**), $[\{(R)$-PE^{Dipp}PIA$\}_2$Ca] (**12**) (Scheme 3.2.3).

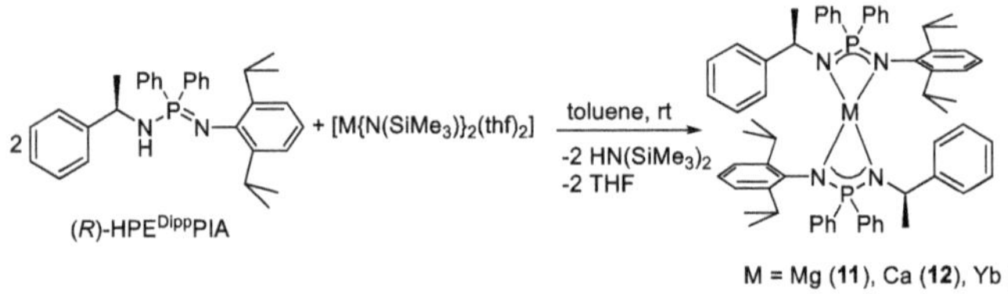

2 (*R*)-HPE^{Dipp}PIA + $[M\{N(SiMe_3)_2\}_2(thf)_2]$ —toluene, rt; -2 $HN(SiMe_3)_2$; -2 THF→

M = Mg (**11**), Ca (**12**), Yb (**13**)

Scheme 3.2.3 Synthesis of complex $[\{(R)$-PE^{Dipp}PIA$\}_2$Mg] (**11**) and $[\{(R)$-PE^{Dipp}PIA$\}_2$Ca] (**12**) and $[\{(R)$-PE^{Dipp}PIA$\}_2$Yb] (**13**).

In the ^1H NMR spectrum of complex **12**, characteristic doublet of quartets for Ph(C*H*)CH_3 and a doublet for Ph(CH)CH_3 is observed at δ= 3.61 ppm and δ= 1.59 ppm, respectively. However, the corresponding signal of the Dipp groups show broad resonances, which might be the result of a hindered rotation. The ^1H NMR spectrum of complex **11** is also apparent to the proposed structure. Furthermore, the $^{31}P\{^1H\}$ NMR spectrum shows resonances at δ = 28.6 ppm (**11**) and δ = 22.6 ppm (**12**) which show significant downfield shift as compared to the ligand (*R*)-HPE^{Dipp}PIA (δ= -10.9 ppm).

The X-ray diffraction analysis of the homoleptic magnesium (**11**) and calcium complex (**12**) was carried out to establish their solid state structures. Both complexes **11** and **12** crystallize in the chiral orthorhombic space group $P2_12_12_1$ (Figure 3.2.4). The molecular structures of complexes **11** and **12** are isostructural. Comparable to **9** and **10**, complexes **11** and **12** also adopt a distorted tetrahedral geometry with the central metal atoms surrounded by four nitrogen atoms of the ligand. With respect to central metal atom (Mg (**11**), Ca (**12**)), the Dipp substituents are trans to each other, which can be attributed to a minimal steric repulsion. Furthermore, the central metal atoms form two N_2PM planes with the NPN backbones, in which the metals are not exactly co-planar with the NPN backbone, but show a rather small deviation with dihedral angles of 1.53(2)° (**11**) and 7.46(2)° (**12**) between NPN and the NMN plane.

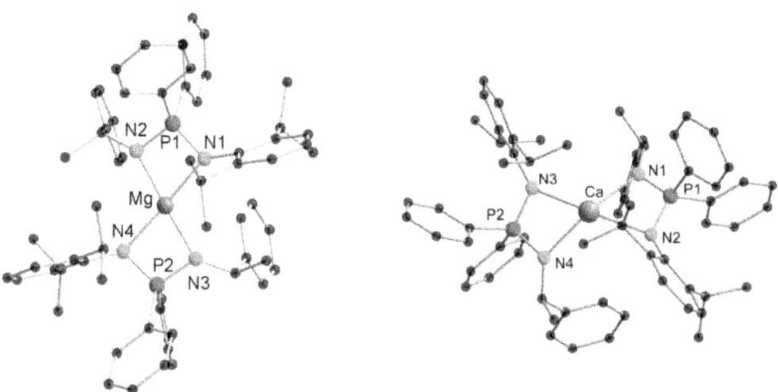

Figure 3.2.4 Molecular structure of complex **11** (left) and **12** (right) in the solid state. All hydrogen atoms are omitted for clarity. Selected bond lengths (Å) and bond angles [°]: For **11**; Mg-N1 2.139(3), Mg-N2 2.059(3), Mg-N3 2.124(3), Mg-N4 2.072(3), P1-Mg 2.722(2), P2-Mg 2.715(2), P1-N2 1.609(3), P1-N1 1.617(3), P2-N3 1.617(3), P2-N4 1.609(3); P1-Mg-P2 172.82(6), N1-Mg-N2 72.51(12), N3-Mg-N4 72.74(12), N2-Mg-N3 106.41(13), N1-Mg-N3 174.15(13), N1-Mg-N4 109.90(13), N2-Mg-N4 164.55(14), N1-P1-N2 100.7(2), N3-P2-N4 100.9(2). For **12**; Ca-N1 2.326(5), Ca-N2 2.437(5), Ca-N3 2.419(5), Ca-N4 2.339(5), Ca-P1 2.995(2), Ca-P2 2.993(2), P1-N1 1.595(5), P1-N2 1.607(5), P2-N3 1.596(5), P2-N4 1.615(5); P1-Ca-P2 159.57(6), N1-Ca-N2 63.6(2), N3-Ca-N4 64.6(2), N2-Ca-N4 116.0(2), N1-Ca-N3 109.1(2), N1-Ca-N4 149.7(2), N2-Ca-N3 167.5(2), N1-P1-N2 103.4(3), N3-P2-N4 104.7(3).

This deviation is comparably higher in the similar tetracoordinated homoleptic complex $[Mg\{Ph_2P(Me_3SiN)_2\}_2]$ (14.1°) and pentacoordinated calcium complex $[(thf)Ca\{Ph_2P(Me_3SiN)_2\}_2]$ (36.4°) possibly due to higher steric constraints of the $(Ph_2P(Me_3SiN)_2)_2^-$.[33] Furthermore, the average NMN bite angle for complex **11** (72.62(12)°) is significantly wider than that observed for complex **12** (64.1(6)°).

The average M-N bond lengths 2.0985(3) Å (M = Mg, **11**) are smaller than 2.380(5) Å for **12**. Obviously, the larger M-N bond length in case of complex **12** is due to the larger ionic radius of calcium compared to magnesium. Additionally, in complex **11**, the P-N bond lengths lie in the range of (1.609(3), 1.617(3) Å), whereas in complex **12**, these values fall in the range of (1.595(5), 1.615(5) Å), which is in between a single and a double P-N bond, in accordance with the complete delocalization of charge over the NPN backbone.

Owing to similar ionic radii of Yb^{+2} (1.02 Å) and Ca^{+2} (1.00 Å), both show similar chemical properties.[72b] Considering this fact, the coordination behaviour of (*R*)-HPE^{Dipp}PIA with divalent Yb^{+2} was investigated. Following the similar amine elimination route as described earlier (*vide supra*), formation of the expected homoleptic complex [{(*R*)-PE^{Dipp}PIA}$_2$Yb] (**13**) was observed (Scheme 3.2.3). The identity of the complex **13** was further established by various analytical techniques such as multinuclear NMR spectroscopy, IR, elemental analysis and also by single crystal X-ray diffraction studies. In the ^{1}H NMR spectrum of **13**, the characteristic resonance of the amine protons N*H* at δ = 2.81 ppm is missing and indicates successful deprotonation. Furthermore, resonances corresponding to Ph(C*H*)CH_3 and Ph(CH)CH_3 appear as a doublet of quartet and a doublet at δ = 4.01 ppm and at δ = 1.84 ppm, respectively. The ^{31}P{^{1}H} NMR spectrum of **13** exhibits a single resonance at δ = 20.6 ppm, which is significantly shifted downfield compared to its precursor ligand (*R*)-HPE^{Dipp}PIA (δ = -10.9 ppm). Complete deprotonation of (*R*)-HPE^{Dipp}PIA was further confirmed by disappearance of NH (ν = 3347 cm^{-1}) stretching frequency in the IR spectrum of complex **13**.

Single crystals of complex **13** suitable for X-ray analysis were obtained from a saturated *n*-pentane solution at low temperature. The molecular structure of **13** suggest that the ytterbium metal shows both a bridging as well as a chelating behaviour with {(*R*)-PE^{Dipp}PIA}$^-$ through (μ-κ^2:κ^2) bonding mode. Similar to **11** and **12**, complex **13** also adopts distorted tetrahedral geometry (Figure 3.2.5). In complex **13**, the Yb-N contact lengths lie in the range of (2.345(4), 2.490(4) Å), which is in consistence to related compounds known in literature.[128] In the Yb-N-P-N planes, the ytterbium centre shows a deviation from the NPN backbones of the ligands with dihedral angles of (17.45(2)°) (N1-P1-N2 *vs* N1-Yb-N2 plane) and (10.63(2)°) (N3-P2-N4 *vs* N3-Yb-N4 plane).

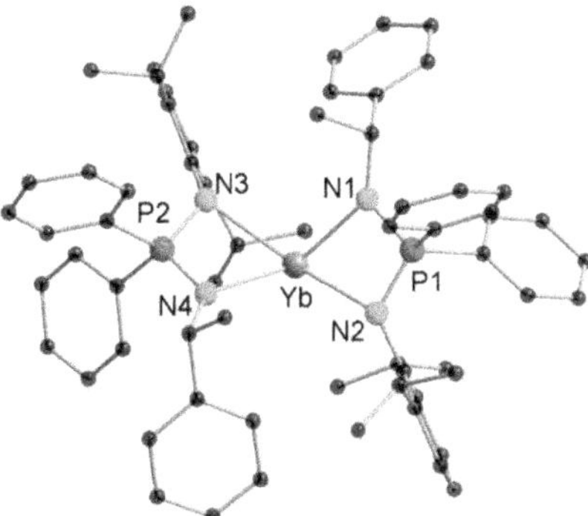

Figure 3.2.5 Molecular structure of **13** in the solid state. All hydrogen atoms are omitted for clarity. Selected bond lengths (Å) and bond angles [°]: Yb-N1 2.345(4), Yb-N2 2.477(4), Yb-N3 2.490(4), Yb-N4 2.352(4), Yb-P1 3.0215(14), Yb-P2 3.0254(14), P1-N1 1.596(4), P1-N2 1.605(5), P2-N3 1.601(5), P2-N4 1.594(4); P1-Yb-P2 143.29(4), N1-Yb-N2 63.2(2), N3-Yb-N4 62.42(13), N2-Yb-N4 106.25(14), N1-Yb-N3 105.77(14), N1-Yb-N4 125.2(2), N2-Yb-N3 157.34(13), N1-P1-N2 104.4(2), N3-P2-N4 103.7(2).

This could also be seen with the nonlinear arrangement of P1-Yb-P2 (143.29(4)°) vector. In complex **13**, the Dipp substituents of the ligand are trans to each other, which is also observed for complex **11** and **12**. The distortion of complex **13** from ideal tetrahedral geometry is attributed to the acute bite angles (N1-Yb-N2 63.2(2)° and N3-Yb-N4 62.42(13)°).

Further, using the same deprotonation method as described above, the reaction of (*R*)-HNEPIA with [Ca{N(SiMe$_3$)$_2$}$_2$(thf)$_2$] in a 2:1 ratio afforded the homoleptic complex [{(*R*)-NEPIA}$_2$Ca] (**14**) (Scheme 3.2.4).

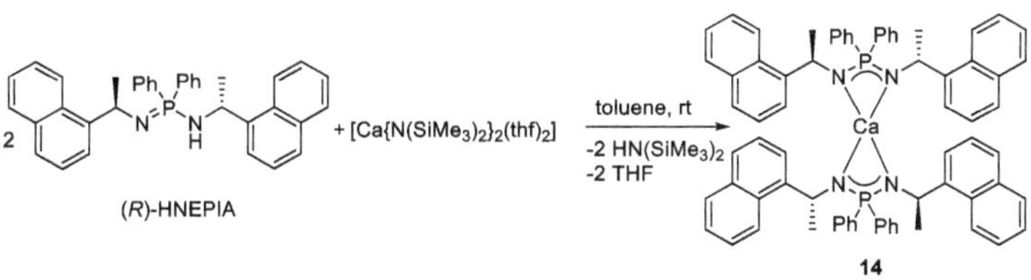

2 (*R*)-HNEPIA + [Ca{N(SiMe$_3$)$_2$}$_2$(thf)$_2$] —(toluene, rt; -2 HN(SiMe$_3$)$_2$; -2 THF)→ **14**

Scheme 3.2.4 Synthesis of complex [{(*R*)-NEPIA}$_2$Ca] (**14**).

Appearance of a single set of resonances in the ^{1}H NMR spectrum of **14** suggests its symmetric nature in solution. Characteristic doublet at δ = 1.39 ppm and a doublet of quartet at δ = 4.82 ppm is assigned to the naph(CH)C*H*$_3$ and naph(C*H*)CH$_3$ moieties, respectively. Corresponding signals in the ^{13}C{^{1}H} NMR spectrum are detected at δ = 28.9 ppm (naph(CH)*C*H$_3$) and at δ = 49.7 ppm (naph(*C*H)CH$_3$). The purity of complex **14** was confirmed by the presence of a singlet at δ = 27.3 ppm, which is significantly shifted downfield as compared to δ = 3.6 ppm for the ligand. Deprotonation of (*R*)-HNEPIA was further confirmed by the absence of the N-H (ν = 3372 cm^{-1}) stretching frequency in the IR spectrum of **14**.

Complex **14** crystallizes in chiral monoclinic space group $P2_1$ with one molecule in the asymmetric unit (Figure 3.2.6).

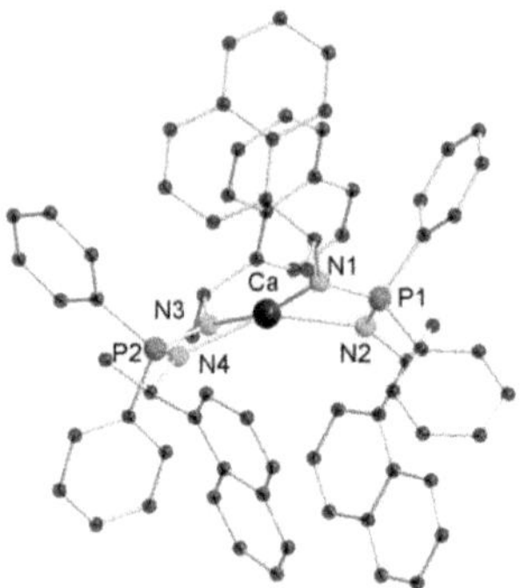

Figure 3.2.6 Molecular structure of **14** in the solid state. All hydrogen atoms are omitted for clarity. Selected bond lengths (Å) and bond angles [°]: Ca-N1 2.390(2), Ca-N2 2.370(2), Ca-N3 2.394(2), Ca-N4 2.374(2), Ca-P1 3.0033(9), Ca-P2 3.0106(8), P1-N1 1.599(2), P1-N2 1.601(2), P2-N3 1.604(2), P2-N4 1.607(2); P1-Ca-P2 170.35(2), N1-Ca-N2 63.97(6), N3-Ca-N4 63.90(7), N1-Ca-N3 167.91(7), N2-Ca-N4 147.51(7), N1-Ca-N4 118.11(7), N1-P1-N2 104.01(10), N3-P2-N4 103.61(10).

No additional coordinating solvent could be seen in the coordination sphere even after performing the reaction in coordinating solvent such as THF. This highlights that, in the absence of coordinating solvents the iminophosphonamide ligand $\{(R)\text{-NEPIA}\}^-$ is sufficiently sterically demanding to stabilize homoleptic species. The solid state structure of **14** shows that calcium atom is encapsulated in a four-membered non-planar CaN_2P metallacycle and a κ^2 chelating mode with two ligand units. The bond angles at the calcium centre range from 63.97(6)° to 170.35(2)° with a considerable distortion from tetrahedral geometry. The naphthyl groups are arranged in a way, that the steric congestion between them is minimized. The Ca centre is slightly above the NPN plane with an angle P1-Ca-P2 equal to 170.35(2)°. Further, the four Ca-N bond lengths in complex **14** lie within the range of (2.370(2), 2.394(2)) Å, which is in a comparable range of (2.344(2), 2.437(5) Å) observed for complex **9** and **12** and also in consistency with related compounds in the literature.[127] The P-N bond lengths in complex **14** are almost equal and lie in the range of (1.599(2), 1.607(2) Å), which again shows that the anionic charge is delocalized all over the NPN backbone.

The bite angles of (R)-HNEPIA in complex **14** (N2-Ca-N1 63.97(6)°, N4-Ca-N3 63.90(7)°) are similar to the analogous complexes **9** and **12**. Furthermore, the NPN bond angle of ligand (*R*)-HNEPIA (119.6(2)°) significantly decreases upon metal coordination (N1-P1-N2 104.01(10)°, N3-P2-N4 103.61(10)°) observed in complex **12**.

3.2.2 Catalysis

The hydroboration of carbonyl compounds involving alkaline earth metals as catalysts is important due to their environmentally friendly nature and from an economic point of view. Known catalysts based on alkaline earth metal complexes include mostly heteroleptic complexes.[92a,92b,121b,121d,129] This is, because homoleptic complexes of alkaline earth metals are considered catalytically silent due to the saturation of their coordination sites. Nevertheless, our group recently reported chiral benzamidinate stabilized homoleptic alkaline earth metal complexes and further showed their catalytic activity in hydrophosphination of styrene.[130] Furthermore, Sen and co-workers reported on hexacoordinated alkaline earth metal (Mg, Ca) complexes based on methylpyridinato β-diketiminate ligands and utilized them as hydroboration agents for carbonyl compounds.[92d]

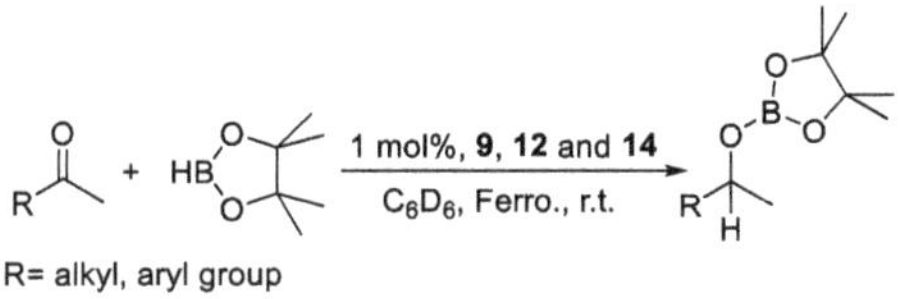

1 mol%, **9**, **12** and **14**

C_6D_6, Ferro., r.t.

R= alkyl, aryl group

Scheme 3.2.5 Catalytic intermolecular hydroboration of ketone

Encouraged by these reports, the homoleptic calcium complexes [{(*R*)-PEPIA}$_2$Ca] (**9**), [{(*R*)-PE^{Dipp}PIA}$_2$Ca] (**12**) and [{(*R*)-NEPIA}$_2$Ca] (**14**) were investigated in hydroboration reactions of ketones. For a preliminary test, an NMR scale (inert condition) reaction between acetophenone and HBpin, using complex **9** as catalyst, was conducted. Upon monitoring the ^{1}H and ^{11}B NMR spectra, a quantitative conversion at room temperature within 5 minutes could be observed. By applying a catalyst loading of 1 mol% and ferrocene as an internal standard, 99% conversion into the corresponding alkoxypinacolboronate ester was achieved. Inspired from this, a variety of para-substituted acetophenones exhibiting electron withdrawing substituents (Br (b), NO_2 (c)) and electron donating substituents (NH_2 (d), CH_3 (e), and OCH_3 (f)) and an aliphatic ketone^g were investigated (Table 3.2.1).

Table 3.2.1 Hydroboration of ketone using catalysts **9**, **12** and **14**.

Entry	Substrate	Product	Catalyst	Yield
a	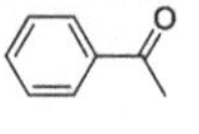 O	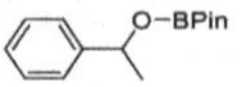 O–BPin	9 12 14	99%
b	Br, O	Br, O–BPin	9 12 14	99%
c	O_2N, O	O_2N, O–BPin	9 12 14	99% 96% 99%
d	H_2N, O	H_2N, O–BPin	9 12 14	99%
e	H_3C, O	H_3C, O–BPin	9 12 14	99%
f	H_3CO, O	H_3CO, O–BPin	9 12 14	99% 97% 99%
g	H_3C, Cl, O	H_3C, Cl, O–BPin	9 12 14	99%

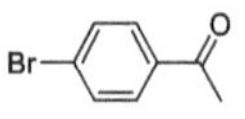

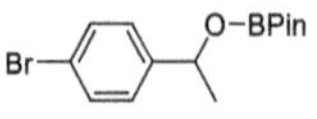

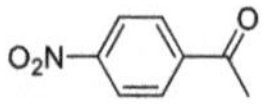

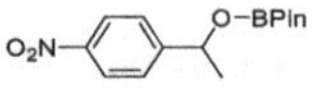

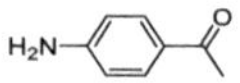

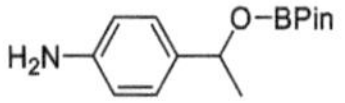

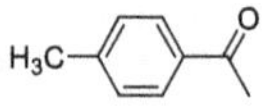

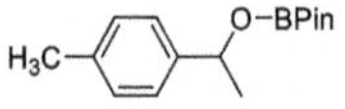

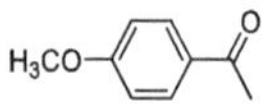

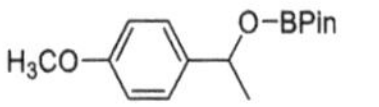

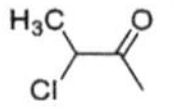

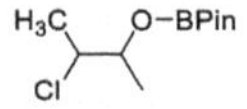

Conditions: Catalysts **9**, **12** and **14**, 1 mol%, C_6D_6, yields are calculated based on ^{1}H-NMR spectroscopy, using ferrocene as the internal standard.

The conversion for almost all these catalysts is observed as 99% within 5 minutes, using a catalyst loading of 1 mol%. The excellent catalytic activity of these complexes (**9**, **12** and **14**) is possibly due to their coordinatively unsaturated nature. Although all these catalysts (**9**, **12** and **14**) are enantiopure, no enantiomeric excess could be obtained in the products, even after doing the reaction at lower temperature.

3.2.3 Photoluminescence studies

The photoluminescence measurements of complexes **9**, **12** and **14** were performed by Dr. Thomas J. Feuerstein and Dr. Sergei Lebedkin (Prof. Manfred M. Kappes working group).

In our previous study, it was shown that the (*R*)-HPEPIA ligand and its alkali metal complexes [{(*R*)-PEPIA}$_2$M$_2$] (M = Li, Na, K, Rb and Cs) show interesting photoluminescence behaviour.[103b] Inspired from this, a similar correlation was anticipated for the alkaline earth metal complexes. For a comparative study, calcium complexes **9**, **12** and **14** were selected to investigate the effect of the nitrogen substituents and the respective orientation of ligand on the PL behaviour.

The study reveals that upon UV irradiation of calcium complexes **9**, **12** and **14** at ~360 nm, blue green photoluminescence is observed.

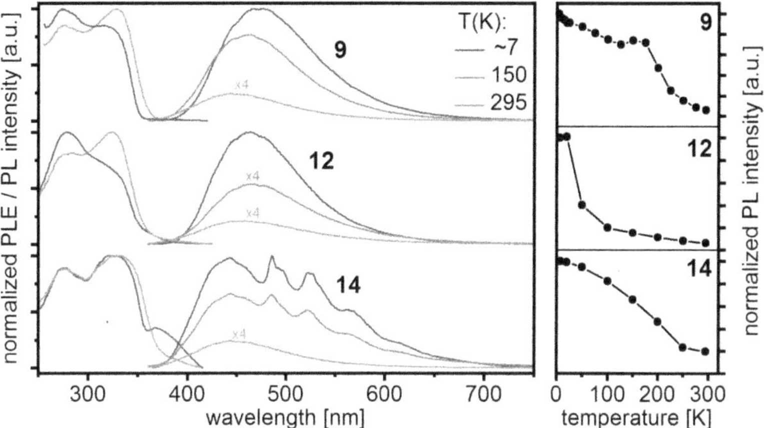

Figure 3.2.7 Left: photoluminescence excitation (PLE) and emission (PL) spectra of compounds **9, 12, 14** at 7, 150 and 295 K. The spectra were recorded/ excited at 480/ 330 nm, respectively. Right: integrated PL intensities plotted against the temperature in the range of 7 to 295 K. For clarity the PL spectra of **9, 12, 14** at 295 K and of **12** at 150 K have been fourfold magnified.

Further, corresponding PL spectra were recorded at 7, 150 and 295 K (Figure 3.2.7). At 295 K, the broad emission bands of complexes **9**, **12** and **14** are centred at 450, 456 and 442 nm. These broad emission bands at 295 K have the width of about 108, 122 and 101 nm (FWHM), respectively. Upon lowering the temperature to 7 K, the emission bands are red shifted to 470 nm (**9**), and 462 nm (**12**), respectively. For complex **14**, four emission bands were observed, which are centred at 442, 486, 524, and 566 nm. These emission bands are attributed to vibronically structured emissions, most likely triggered by the naphthyl substituents in the ligand backbone. Correspondingly, the PL quantum yield of complex **12** is relatively low (<1%) in comparison to complex **9** (22%) and **14** (7%) under ambient conditions (Table 3.2.2). In case of the calcium complex **9**, the PL behaviour is not consistent to that observed for the dimeric alkali metal complexes of the same ligand (*R*)-HPEPIA. As shown in figure 3.2.7 (right), the integrated PL intensity of complex **9** below ~100 K show a long-lived phosphorescence with a lifetime of ca. 5.6 msec, which on increasing temperature to 295 K changes into a faster decay with an effective lifetime of ca. 24 μsec. The transition from slower decay into faster decay occurs at the temperature ranging from 130 to 180 K, which correlates to a non-monotonic dependence of the integrated PL intensity vs. temperature (Figure 3.2.7).

This kind of hump in the integrated PL intensity of complex **9** was also observed in the dimeric alkali metal complexes [{(*R*)-PEPIA}$_2$Na$_2$], [{(*R*)-PEPIA}$_2$K$_2$] and [{(*R*)-PEPIA}$_2$Rb$_2$].

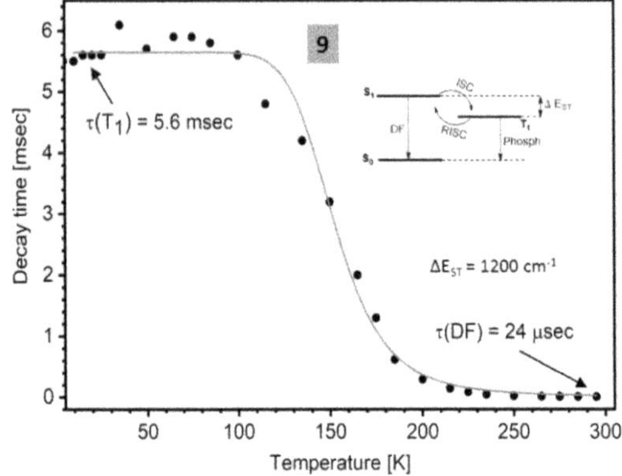

Figure 3.2.8 PL decay time of solid complex **9** vs. temperature. The PL was excited with a nsec-pulsed laser at 337 nm, recorded at 480 nm and fit with monoexponentially curves. The red curve depicts the fit according to (equation 1, section 3.1.4). The fit yielded the energy separation ΔE between S_1 and T_1 states of about 1200 cm^{-1}. Below T ~100 K the emission is predominantly phosphorescence with the lifetime of 5.6 msec. Increasing the temperature activates the TADF mechanism, resulting in the effective PL lifetime of ca. 24 μsec at room temperature.

Based on a simple TADF model of thermally equilibrated S_1 and T_1 states (equation 1),[131] the estimated energy separation ΔE (energy separation between S_1 and T_1) is ca. 1200 cm^{-1} (148 meV) (Figure 3.2.8). This value of ΔE is comparatively higher than those observed for the dimeric alkali metal complexes (~600-750 cm^{-1}). In comparison to **9**, the calcium complex **12** shows a very long-lived phosphorescence components (τ = 30 msec at 5 K) and minor fast fluorescence below 20 K. The transition corresponding to the TADF occurs in between ~20 and ~75 K, indicating a much smaller energy gap ΔE between S_1 and T_1 states, roughly estimated as ~330 cm^{-1}. This value of ΔE is much smaller in comparison to those of complex **9** and other alkali metal complexes. As from the integrated PL spectra of **12**, the fast (fluorescence) component becomes dominant above ~200 K. These ambiguities in the PL behaviour of complexes **9** and **12** could be attributed to their structural difference. In case of complex **9**, the calcium metal is exactly coplanar with the NPN backbone, whereas in complex **12**, due to steric repulsion of Dipp substituents, it lies slightly above to the NPN backbone, having an average dihedral angle of 7.46(2)° between the NPN and NMN plane. This kind of correlation between the structural and photophysical properties were observed in the alkali metal complexes as well.[55]

Table 3.2.2 Characteristic spectroscopic parameters for the photoluminescence (PL) of complexes **9**, **12** and **14**.

Entry	Compound		9	12	14
1	**Nitrogen substituent**		1-methylbenzyl	Dipp	Naphthyl
2	**λ_{PL} (7 K)[a]**	**[nm]**	470	462	442(486/524/566)[g]
3	**λ_{PL} (295 K)[b]**	**[nm]**	450	456	442
4	**FWHM(7 K)[c]**	**[nm]**	115	104	163
5	**FWHM(295 K)**	**[nm]**	108	122	101
6	**$\lambda_{PLE}(\varphi)$[d]**	**[nm]**	330	330	330
7	**φ(295 K)[e]**	**[%]**	22	<1	7
8	**τ_1(7 K)**	**[ns]**	<10	<10	<10
9	**τ_2(Phosph, 7 K)[f]**	**[ms]**	5.5	30	–
10	**τ_2(DF, 295 K)[f]**	**[µs]**	1.9	0.2	–

[a] Excitation and emission wavelengths applied for recording the PL and PLE spectra; [b] Band maxima (nm) in emission and excitation spectra; [c] Bandwidth of the PL emission; [d] Excitation wavelength used for determination of the PL quantum yield at ambient temperature; [e] PL quantum yield φ(295 K) determined according to ref.[112]; [f] Long-lived components of the PL decay of **9** and **12** well follow monoexponential curves; [g] The values in brackets concern the vibronically structure emission which was observed for compound **14** at low temperatures.

In contrast to TADF observed in case of complex **9** and **12**, only a fast PL component with a decay time < 10 nsec (time resolution of the apparatus) could be detected for complex **14** both at low temperatures as well as under ambient conditions. This can be assigned to the fluorescence of the naphthyl moieties in **14**. Its efficiency amounts to 7% at room temperature. For complex **14**, by rising the temperature over the range of 7-295 K, a substantial decrease in the PL efficiency was observed comparable to **9** and **12** (Figure 3.2.8).

3.3 Copper and zinc complexes of iminophosphonamides

3.3.1 Introduction

Over the past decade, luminescent transition metal complexes with d^{10} metal ions have been widely explored and applied in a variety of photo-functional materials.[132] The major advantage of these metal complexes lies in the lack of metal centred charge transfer states. Therefore, no alteration in the luminescent excited states occurs by thermal equilibrium.[133] In this regard, monovalent group 11 "coinage" metals and Zn(II) complexes are of great interest.[132,134] Utilizing suitable ligand systems, the resulting dinuclear coinage metal complexes exhibit interesting metallophilic interactions, in which the metal centres are in closer proximity than the sum of their van der Waal radii. In this context, a large variety of neutral and anionic ligands have been screened to prepare light-emitting Cu(I), Ag(I) and Au(I) complexes with different nuclearity.[8b,132] From an economic point of view, luminescent complexes involving copper and zinc are of specific interest.

In particular, the luminescence properties of copper(I) complexes have been largely studied with a variety of *N*-chelating ligands, such as bisimines, substituted 1,10-phenanthrolines or amidinates.[135] A variety of copper(I) complexes have also been tested for potential OLED applications.[136] Zn(II) containing metal complexes with Schiff base ligands have also attracted a great interest due to their strong photoluminescence emissions in the blue region of the visible spectrum.[137]

Inspired by the results obtained from the PL study of the alkali and alkaline earth metal complexes (sections 3.1.3.5 and 3.2.3), the PL behaviour of the respective zinc and copper complexes has been investigated. Herein, a variety of substituted iminophosphonamine ligand systems has been screened (Chart 3.3.1).

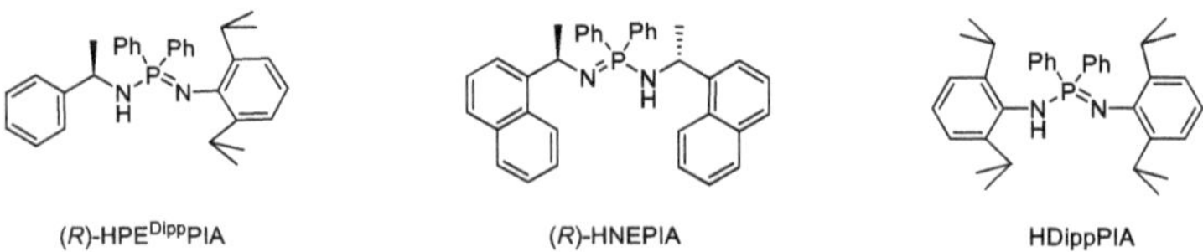

Chart 3.3.1 Iminophosphonamine ligands used in the synthesis of Cu(I) and Zn(II) complexes.

3.3.2 Synthesis of Cu(I) and Zn(II) iminophosphonamide complexes

Cu(I) and Zn(II) complexes in association with the (*R*)-HPEPIA ligand have already been discussed in previous studies from our group.[138] Herein, the coordination chemistry of related ligand systems (see Chart 3.3.1) is investigated.

In order to introduce the unsymmetrical {(*R*)-PE^{Dipp}PIA}$^{-}$ ligand into the coordination sphere of Cu(I), a one pot reaction was performed between CuCl, (*R*)-HPE^{Dipp}PIA and $KN(SiMe_3)_2$ in an equimolar ratio. As a result, facile elimination of KCl and formation of the expected copper complex, [{(*R*)-PE^{Dipp}PIA}Cu]$_2$ (**15**), occurred (Scheme 3.3.1).

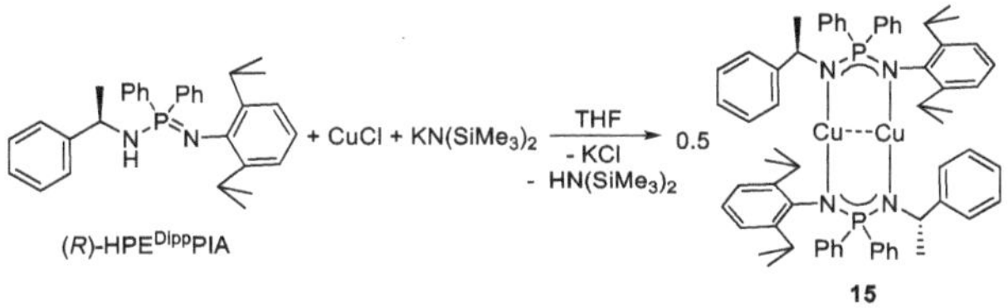

Scheme 3.3.1 Synthesis of the Cu(I) complex [{(*R*)-PE^{Dipp}PIA}Cu]$_2$ (**15**).

The formation of complex **15** was further confirmed by various analytical techniques such as multinuclear NMR, IR, and elemental analysis. In the $^{31}P\{^{1}H\}$ NMR spectrum of **15**, one single resonance at δ = 38.0 ppm is observed, which is significantly shifted downfield as compared to δ = -10.9 ppm for (*R*)-HPE^{Dipp}PIA. Furthermore, the resonances in the ^{1}H NMR spectrum are consistent with the proposed structure. The elemental analysis of the product is in good agreement with the molecular formula as well.

The solid-state structure of complex **15** was determined by single crystal X-ray diffraction studies. Complex **15** crystallizes in the orthorhombic chiral space group $P2_12_12_1$ with one molecule in the asymmetric unit (Figure 3.3.1). The molecular structure shows that two copper centres are bridged by the iminophosphonamide ligands, in a structure similar to that of the related [{$Ph_2P(NSiMe_3)_2$}$_2Cu_2$] complex.[58] The Cu-Cu bond length in **15** is 2.53(7) Å, falling in the range of cuprophilic interactions,[59] and is comparable to the intermetallic distance of 2.53(2) Å observed in [{(*R*)-PEPIA}Cu]$_2$. It is however slightly shorter than that observed for [{$Ph_2P(NSiMe_3)_2$}$_2Cu_2$] (2.63(1) Å). The two flanking Dipp groups in complex **15** are *trans* to each other, probably to minimize the steric repulsion. The Cu-N bond lengths are almost identical and lie in the range of previously reported complexes.[139]

Furthermore, the P-N bond lengths are in between a single and a double bond, suggesting delocalisation of the anionic charge over the NPN framework.

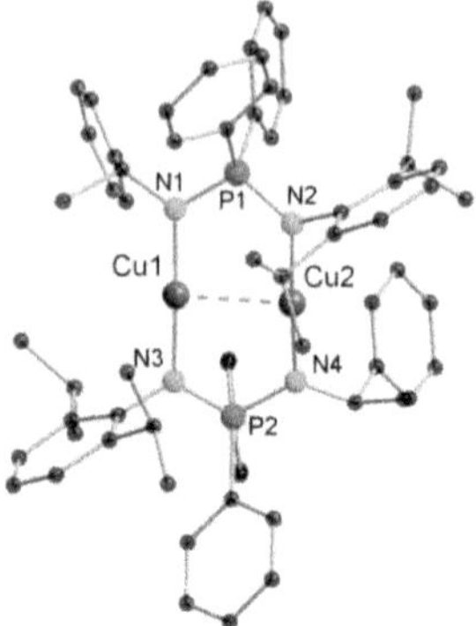

Figure 3.3.1 Molecular structure of **15** in the solid state. All hydrogen atoms are omitted for clarity. Selected bond lengths (Å) and bond angles [°] Cu1···Cu2 2.5303(7), Cu1-N1 1.866(3), Cu2-N2 1.870(3), Cu1-N3 1.867(3), Cu2-N4 1.866(3), P1-N1 1.608(3), P1-N2 1.611(3), P2-N3 1.609(3), P2-N4 1.613(3); N1-Cu1-N3 177.7(2), N2-Cu2-N4 177.4(2), N1-Cu1-Cu2 92.68(10), N2-Cu2-Cu1 89.33(10), N3-Cu1-Cu2 89.22(10), N4-Cu2-Cu1 92.84(11), P1-N1-Cu1 121.2(2), P1-N2-Cu2 122.1(2), P2-N3-Cu1 122.4(2), P2-N4-Cu2 121.0(2), N1-P1-N2 109.8(2), N3-P2-N4 109.7(2); P-P distance [Å]: P1-P2 5.501(2); NPPN Torsion angle [°]: N1-P1-P2-N4 179.8(2).

In order to synthesize Zn(II) complexes with the (*R*)-PE^{Dipp}PIA ligand, the deprotonation and metallation of (*R*)-HPE^{Dipp}PIA was performed by treatment with $ZnPh_2$ in a molar ratio of 2:1 which resulted in the formation of [{(*R*)-PE^{Dipp}PIA}$_2$Zn] (**16**) (Scheme 3.3.2).

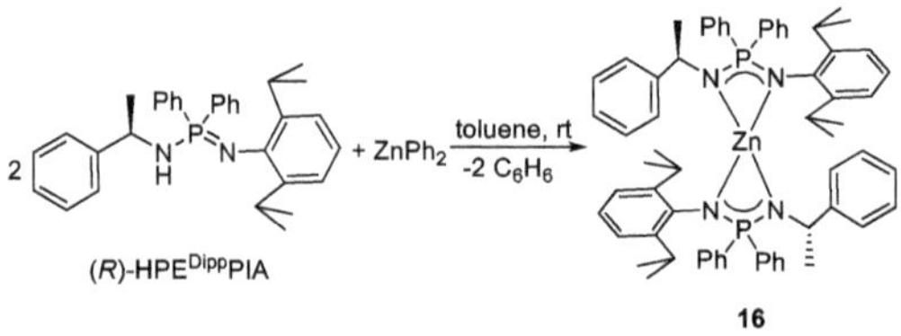

2 (*R*)-HPE^{Dipp}PIA + $ZnPh_2$ —(toluene, rt; $-2\ C_6H_6$)→ **16**

Scheme 3.3.2 Synthesis of [{(*R*)-PE^{Dipp}PIA}$_2$Zn] (**16**).

Successful formation of complex **16** was indicated by disappearance of the N*H* resonance in the ^{1}H NMR spectrum. It should be noted that the ^{1}H NMR spectrum of **16** shows a single set of resonances for the different signals, suggesting a certain degree of symmetry in solution. The Ph(C*H*)CH$_3$ signal appears as a doublet of a quartets in the aliphatic region at δ= 3.79 ppm ($^3J_{PH}$ = 18.57, $^3J_{HH}$ = 6.53 Hz). A doublet at δ = 1.37 ppm ($^3J_{HH}$ = 6.52 Hz) is assigned to the Ph(CH)C*H*$_3$ protons. The purity of complex **16** was further confirmed by the detection of only one peak at δ= 37.5 ppm in the ^{31}P{^{1}H} NMR spectrum, which is significantly shifted downfield compared to δ = -10.9 ppm for (*R*)-HPE^{Dipp}PIA. Furthermore, complete deprotonation of (*R*)-HPE^{Dipp}PIA was confirmed by disappearance of the NH stretching band in the IR spectrum of **16**.

The molecular structure of **16** was determined by single crystal X-ray analysis. Suitable crystals were obtained by slow evaporation of a *n*-pentane solution. The complex crystallizes in the orthorhombic space group $P2_12_12_1$ with one molecule in the asymmetric unit (Figure 3.3.2). The Zn metal is coordinated by four nitrogen atoms of two {(*R*)-PE^{Dipp}PIA}$^-$ ligands. The central zinc atom is part of two four-membered non-planar ZnN_2P metallacycles and exhibits a slight deviation from the N1-P1-N2 and N3-P2-N4 planes with dihedral angles of 0.56(2)° and 2.47(2)°, respectively. In order to minimize the steric repulsion, the Dipp-groups adopt *trans* positions with respect to central metal. Owing to the unsymmetrical nature of (*R*)-HPE^{Dipp}PIA, the Zn-N bond distances are not identical and lie in the range 1.994(3)-2.197(3) Å. The P-N bond lengths in complex **16** are almost identical (1.610(3), 1.603(3), 1.609(3) and 1.603(3) Å), suggesting complete delocalisation of the anionic charge. The NPPN-torsion angle in complex **16** is 166.33(2)°, which is slightly lower than that in **15** (N1-P1-P2-N4 179.8(2)°).

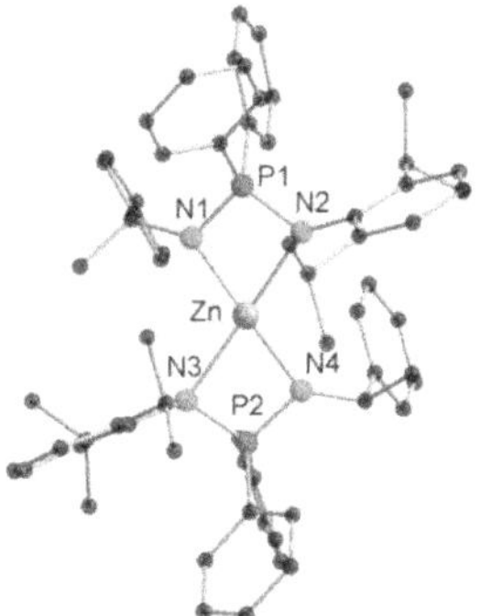

Figure 3.3.2 Molecular structure of **16** in the solid state. All hydrogen atoms are omitted for clarity. Selected bond lengths (Å) and bond angles [°]: Zn-N1 1.994(3), Zn-N2 2.197(3), Zn-N3 2.194(3), Zn-N4 1.994(3) P1-N1 1.610(3), P1-N2 1.603(3), P2-N3 1.609(3), P2-N4 1.603(3); N1-Zn-N4 166.25(13), N1-Zn-N2 72.18(11), N2-Zn-N3 171.05(11), N2-Zn-N4 107.50(11), N1-Zn-N3 110.50(12), N3-Zn-N4 72.03(11), N1-P1-N2 100.7(2), N3-P2-N4 100.4(2), P1-N1-Zn 97.2(2), P1-N2-Zn 89.80(13), P2-N3-Zn 89.92(13), P2-N4-Zn 97.6(2). P-P distance [Å]: P1-P2 5.4302(2); NPPN torsion angle [°]: N1-P1-P2-N4 166.33(2).

After the successful isolation of the binuclear copper complex **15**, the ligand's substituents were altered to increase the π character of the ligand. Indeed, functionalised extended π conjugated systems exhibit interesting photophysical properties,[140] which is due to the possibility of π-π interaction.[141] Therefore, the {(*R*)-NEPIA}$^-$ ligand was chosen and features naphthyl groups at the nitrogen centres. A salt metathesis reaction between [{(*R*)-NEPIA}$_2$K$_2$] (**8**) and CuCl was conducted in a 1:2 molar ratio (Scheme 3.3.3). After extraction and standard work up procedure, complex **17** was isolated as a colourless solid.

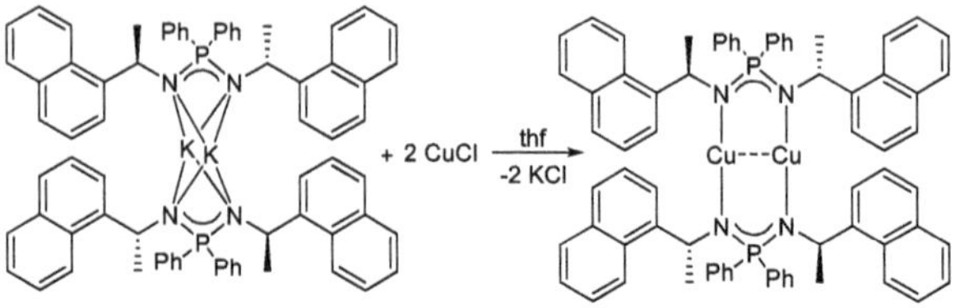

Scheme 3.3.3 Synthesis of [{(*R*)-NEPIA}Cu]$_2$ (**17**).

The ^{1}H NMR spectrum of complex **17** is consistent with a symmetric arrangement in solution. More precisely, the characteristic doublet of quartets at δ = 5.12 ($^3J_{PH}$ = 16.3 Hz, $^3J_{HH}$ = 6.4 Hz) is assigned to the naph(C*H*)CH$_3$ protons, whereas the doublet at δ = 1.9 ppm ($^3J_{HH}$ = 6.5 Hz) is attributed to the naph(CH)C*H*$_3$ protons. In the ^{13}C{^{1}H} NMR spectrum, the methine and methyl moieties can be assigned to the resonances at δ = 50.6 ppm (naph(*C*H)CH$_3$) and δ = 31.3 ppm (naph(CH)*C*H$_3$), respectively. The purity of the complex can be assessed by the detection of a single resonance at δ = 40.2 ppm in the ^{31}P{^{1}H} NMR spectrum, which is significantly shifted downfield as compared to δ = 18.5 ppm in **8**, whereas shifted upfield as compared to δ = 38.0 ppm in the aforementioned dicopper compound **15**.

Complex **17** crystallizes in the orthorhombic chiral space group $P2_12_12_1$ with one molecule in the asymmetric unit (Figure 3.3.3). The Cu1···Cu2 interaction in complex **17** is 2.5054(8) Å, which falls within the range of cuprophilic interactions.[114] The Cu-N bond lengths in **17** are Cu1-N1 1.892(4) Å, Cu1-N4 1.887(4) Å, Cu2-N2 1.880(4) Å and Cu2-N3 1.881(4) Å) slightly longer to those observed in **15**. The N1-Cu1-N4 (175.6(2)°) and N2-Cu2-N3 (174.7(2)°) bond angles are almost linear. Furthermore, the N1-P1-P2-N3 torsion angle of 127.59(2)° in complex **17** is significantly narrower than that in **15** (179.8(2)°). The P-N contact lengths in **17** are almost equal (P1-N1 1.611(4), P1-N2 1.608(4), P2-N3 1.611(4) and P2-N4 1.612(4)), suggesting delocalisation of the anionic charge over the NPN backbone.

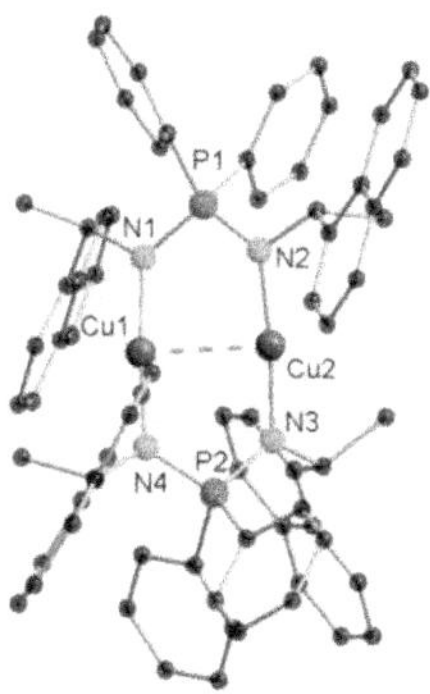

Figure 3.3.3 Molecular structure of **17** in the solid state. All hydrogen atoms are omitted for clarity. Selected bond lengths (Å) and bond angles [°]: Cu1···Cu2 2.5054(8), Cu1-N1 1.892(4), Cu1-N4 1.887(4), Cu2-N2 1.880(4), Cu2-N3 1.881(4), P1-N1 1.611(4), P1-N2 1.608(4), P2-N3 1.611(4), P2-N4 1.612(4); N1-Cu1-Cu2 88.37(13), N1-Cu1-N4 175.6(2), N4-Cu1-Cu2 87.23(13), N2-Cu2-Cu1 86.73(14), N2-Cu2-N3 174.7(2), N3-Cu2-Cu1 88.02(13), N1-P1-N2 109.3(2), N3-P2-N4 109.4(2), P1-N1-Cu1 118.0(2), P1-N2-Cu2 116.7(2), P2-N4-Cu1 119.3(2), P2-N3-Cu2 114.8(2), P-P distance [Å]: P1···P2 5.420(2); NPPN torsion angle [°]: N1-P1-P2-N3 127.59(2).

In order to study the effect of the steric demand of the ligand on the molecular arrangements, the coordination behaviour of copper(I) towards the sterically demanding achiral ligand *P,P*-diphenyl-*N,N'*-bis(2,6-diisopropyl phenyl)phosphinimidic amine (HDippPIA) was investigated. The reaction between CuCl, HDippPIA, and $KN(SiMe_3)_2$ was carried out in THF (Scheme 3.3.4).

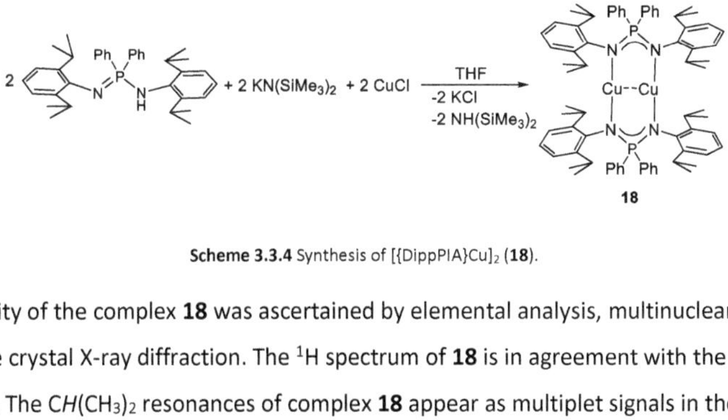

2 HDippPIA + 2 $KN(SiMe_3)_2$ + 2 CuCl —THF, -2 KCl, -2 $NH(SiMe_3)_2$→ **18**

Scheme 3.3.4 Synthesis of $[\{DippPIA\}Cu]_2$ (**18**).

The identity of the complex **18** was ascertained by elemental analysis, multinuclear NMR, IR, and single crystal X-ray diffraction. The ^1H spectrum of **18** is in agreement with the proposed structure. The $CH(CH_3)_2$ resonances of complex **18** appear as multiplet signals in the range of δ = 4.45-3.59 ppm. In contrast, the $CH(CH_3)_2$ resonances appear as discrete doublets with almost equal coupling constants. Furthermore, the $^{31}P\{^1H\}$ NMR spectrum of **18** exhibits a peak at δ = 8.4 ppm, which is significantly shifted downfield as compared to that of free HDippPIA (δ = -13.5 ppm).[26]

Complex **18** crystallizes in the monoclinic space group $P2_1/n$ with half of the molecule in the asymmetric unit (Figure 3.3.4). The structural features of complex **18** are almost similar to those observed for **15** and **17**. The ligand exhibits a bridging μ- κ^1: κ^1 coordination mode with the copper centres. The intermetallic Cu1···Cu1' interaction in **18** is 2.5292(10) Å, which is consistent with that observed in the related compounds [{(2,6-$Me_2C_6H_3N)_2C(H)\}_2Cu_2$][135d] and [{*i*PrNC(Me)N*i*Pr}$_2Cu_2$][17]. The N1-Cu1-N2' angle is of 176.82(12)° which is close to linearity, similarly the arrangements in complexes **15** and **17**. The respective Cu1-N1 and Cu1'-N2 bond lengths of 1.895(3) and 1.886(3) Å in **18** are slightly longer than that observed for [{(2,6-$Me_2C_6H_3N)_2C(H)\}_2Cu_2$] (1.869(1) Å). Furthermore, the N1-P1-P1'-N1' torsion angle of 180.00(2)° in **18**, is significantly wider as compared to 127.59(2)° (**17**), presumably because of the higher steric demand induced by the Dipp-groups.

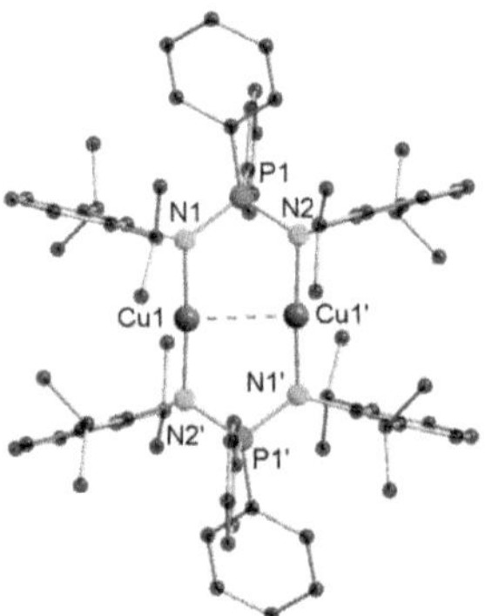

Figure 3.3.4 Molecular structure of **18** in the solid state. All hydrogen atoms are omitted for clarity. Selected bond lengths (Å) and bond angles [°]: Cu1···Cu1' 2.5292(10), Cu1-N1 1.895(3), Cu1'-N2 1.886(3), P1-N1 1.617(3), P1-N2 1.614(3); N1-Cu1-Cu1' 95.85(9), N2-Cu1-Cu1' 86.61(9), N2'-Cu1-N1 176.82(12), N1-P1-N2 107.9(2), Cu1-N1-P1 113.4(2), Cu1-N2-P1 121.3(2), Cu1-N1-P1 113.4(2), P-P distance [Å]: P1-P1' 5.4327(12); NPPN torsion angle [°]: N1-P1-P2-N1' 180.00(2).

In contrast to the synthesis of compound **16**, when HDippPIA was reacted with diphenyl zinc in a 2:1 molar ratio, formation of the heteroleptic complex [{DippPIA}ZnPh] (**19**) was observed instead of the homoleptic analogue. This unexpected reactivity may be due to sterically hindered nature of HDippPIA which prevents the coordination of two ligands on the same Zn(II) metal centre. The yield of complex **19** could be further improved by using an equimolar ratio of diphenyl zinc and HDippPIA (Scheme 3.3.5). Complex **19** exhibits a good solubility in common organic solvents such as THF, toluene and Et_2O, however, it is only slightly soluble in nonpolar solvents such as *n*-pentane.

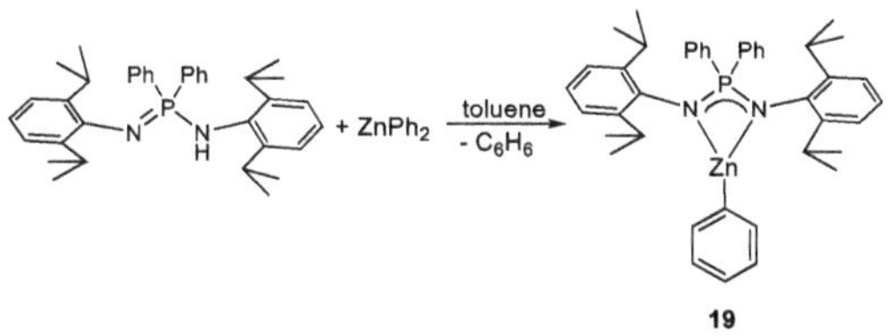

Scheme 3.3.5 Synthesis of [{DippPIA}ZnPh] (**19**).

The ^{1}H NMR spectrum of complex **19** confirms the heteroleptic nature of the complex, which was found stable in solution as no sign of ligand redistribution to homoleptic complexes was observed. The aromatic resonances appear in the range δ = 7.63-6.77 ppm. The characteristic septet of the $CH(CH_3)_2$ moiety appears at δ = 3.88 ppm ($^{3}J_{HH}$ = 6.81 Hz), whereas the $CH(CH_3)_2$ resonance appears as doublet at δ = 1.03 ppm ($^{3}J_{PH}$ = 6.87 Hz). In the $^{13}C\{^{1}H\}$ NMR spectrum, the resonances at δ = 29.4 ppm and δ = 24.1 ppm can be attributed to $CH(CH_3)_2$ and $CH(CH_3)_2$, respectively. In the $^{31}P\{^{1}H\}$ NMR spectrum, the appearance of a single peak at δ = 17.2 ppm further indicates the purity of the product.

The solid state structure of **19** was determined by X-ray diffraction studies. Complex **19** crystallizes in the monoclinic space group *P*2$_1$/*n* with one molecule in the asymmetric unit (Figure 3.3.5). The molecular structure of **19** shows that the zinc centre exists in a monomeric form.

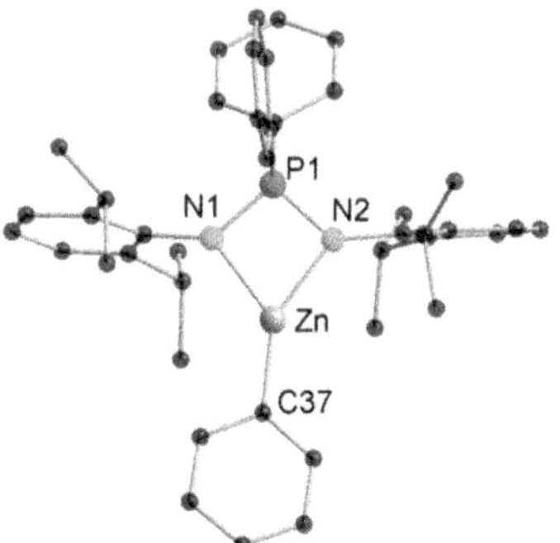

Figure 3.3.5 Molecular structure of **19** in the solid state. All hydrogen atoms are omitted for clarity. Selected bond lengths (Å) and bond angles [°]: Zn-N1 2.026(2), Zn-N2 1.996(2), Zn-C37 1.932(3), P1-N1 1.604(2), P1-N2 1.617(2); N1-Zn-N2 74.51(9), C37-Zn-N1 136.06(11), C37-Zn-N2 148.19(11), N1-P1-N2 98.22(11).

The overall structure of **19** is similar to that of the related [{DippPIA}ZnMe] and [{DippPIA}ZnEt] complexes reported by Stasch and co-workers.[26] The Zn-N (Zn-N1 2.026(2) Å and Zn-N2 1.996(2) Å) and Zn-C (Zn-C37 1.932(3) Å) bond lengths in **19** are consistent with those in related complexes.[23b] The acute N1-Zn-N2 bite angle of 74.51(9)° in complex **19** is wider than that of 65.9(2)° in the related amidinate complex [{MeC(NDipp)$_2$}$_2$Zn]. The P-N bond distance in **19** lies in between a single and a double bond, suggesting delocalisation of the anionic charge over the NPN backbone.[103b]

3.3.3 Photoluminescence study

The photoluminescence properties of complexes **15-19** were examined in collaboration with Dr. Sergei Lebedkin and Dr. Thomas J. Feuerstein.

The characterization of the PL properties of **15-19** was mainly motivated by the results obtained from the alkali metal complexes [{(*R*)-PEPIA}$_2$M$_2$] (M = Li, Na, K, Rb and Cs) and complexes **9**, **12** and **14** showing temperature and structure dependent PL behavior. Additionally, the copper and zinc complexes of (*R*)-HPEPIA discussed before revealed TADF properties.[103a] The idea of PL study was further boosted when examined the structural parameters of the copper complexes **15**, **17** and **18** which display Cu···Cu distances in the expected range for cuprophilic interactions therefore corresponding complexes were anticipated to have interesting PL behaviour. For clarity, the PL of copper and zinc complexes are discussed separately.

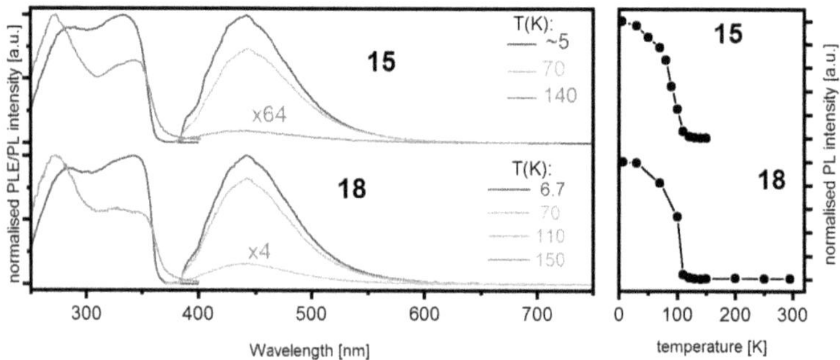

Figure 3.3.6 Left: photoluminescence excitation (PLE) and emission (PL) spectra of complexes **15** and **18** in the temperature range 5-150 K. Right: integrated PL intensities plotted against the temperature in the range of 7 to 295 K. For clarity the PL spectra of **15** has been magnified 64 times at 140 K and of **18** at 150 K has been four-fold magnified.

The PL spectrum of **15** exhibits a broad emission band at different temperatures (Figure 3.3.6). At 4.7 K, an intense band centred at λ_{emis} = 442 nm (λ_{exc} = 333 nm) is observed with a full width at half maximum (FWHM) of 79 nm. Upon increasing the temperature, the emission bands show slight red shifts. The PL intensity of **15** decreases quickly upon increasing the temperature. The temperature dependency of the PL intensity and the decay lifetime of **15** is relatively similar to the alkali metal complexes and the calcium complexes **9** and **12**. The PL intensity of **15** decreases slowly in the temperature range from 4.7 to 70 K whereas above 70 K, it decreases exponentially and attains the steady state in the temperature range from 120-150 K. As shown in the graph of PL decay time *vs* temperature (Figure 3.3.7). At 4.7 K complex **15** shows fast phosphorescence (with $\tau(T_1)$ = 0.22 ms) whereas upon increasing the temperature at around 140 K, it turns into delayed S_1 fluorescence (with the effective time τ = 0.25 μs at 140 K). Although based on equation 1, (section 3.1.3.5), a singlet-triplet energy difference $\Delta E(S_1\text{-}T_1)$ of 1102 cm^{-1} was calculated, which is in the accepted range for TADF (Figure 3.3.7).[133] However the complex is not contributing in any delayed fluorescence suggesting quenching of photoluminescence *via* nonradiative relaxation from excited singlet (*i.e.* S_1) state.

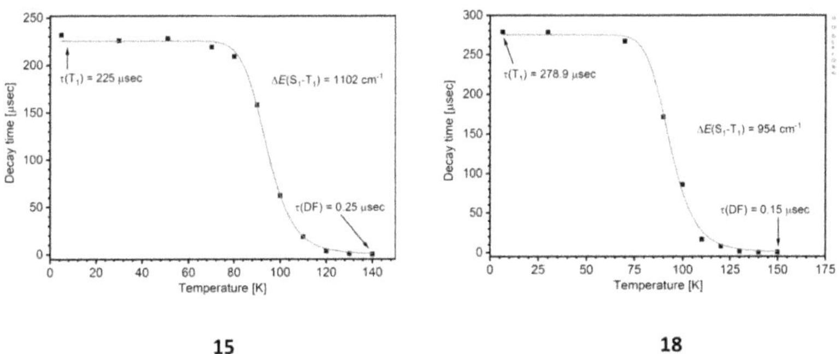

Figure 3.3.7 PL decay time *vs* temperature, left for **15** and right for **18**. For **15**: at low temperatures, the emission is phosphorescence from the T_1 state. The red line shows the fit according to equation (1) (section 3.1.4) yielding $\tau(T_1)$ = 225 μs, $\tau(S_1)$ = 0.25 μs and S_1-T_1 energy separation of 1102 cm^{-1}. For **18**: $\tau(T_1)$ = 279 μs, $\tau(S_1)$ = 0.15 μs and S_1-T_1 energy separation of 954 cm^{-1}.

Next, the Dipp-substituted copper complex **18** was investigated and the PL data was compared with the results for **15**. The PL spectrum of **18** shows a broad emission band centered at 443 nm with a FWHM of 80 nm at 6.7 K, which is not shifted upon raising the temperature to 150 K and widens to a FWHM of 108 nm.

Table 3.3.1 Characteristic spectroscopic parameters for the photoluminescence (PL) of complexes **15**, **17** and **18**.

Entry	Complex	15	17	18
1	**Ligand system**	{(*R*)-PE^{Dipp}PIA}$^-$	{(*R*)-NEPIA}$^-$	{DippPIA}$^-$
2	**λ_{PL}^{a}(LT) [nm]**	442	500	443
3	**λ_{PL}^{b}(HT) [nm]**	448	501	443
4	**FWHM^{c}(LT) [nm]**	79	75	80
5	**FWHM^{d}(HT) [nm]**	78	89	108
6	**τ_1(5 K) [ns]**	<10	<10	<10
7	**τ_2(Phosph)^{f}(LT) [µs]**	225	-	279
8	**τ_2(DF)^{f}(HT) [µs]**	0.2	-	0.2

[a] Excitation and emission wavelengths applied for recording the PL and PLE spectra (recorded at LT = 4.7 K (**15**), 5 K (**17**), 6.7 K (**18**)); [b] Band maxima (nm) in emission and excitation spectra (recorded at HT = 120 K (**15**), 295 K (**17**), 150 K (**18**)); [c] Bandwidth of the PL emission at LT; [d] Bandwidth of the PL emission at HT. [f] Long-lived components of the PL decay of **15** and **18** well follow monoexponential curves.

The PL intensity of these complexes decreases slightly from 6.5 K to 30 K whereas it decreases abruptly upon raising the temperature from 30 to 100 K and attains the steady state in the temperature range of 110 to 140 K. In consistency to **15**, complex **18** also shows phosphorescence at 6.7 K (with $\tau(T_1)$ = 279 µs) which turns into delayed fluorescence at 140 K with the effective decay time of τ = 0.15 µs (Figure 3.3.7). This crossover from phosphorescence to delayed fluorescence occurs at (~ 90-120 K). Using equation 1 (section 3.1.3.5), the calculated energy difference $\Delta E(S_1\text{-}T_1)$ is 954 cm^{-1}, which is slightly lower in comparison to **15** and also in the accepted range for TADF.[133] Similar to **15** since this complex is not contributing in the delayed fluorescence suggesting the nonradiative relaxation is predominant.

Furthermore, in order to examine the influence of an enhanced conjugated π system at the nitrogen substituents on the PL behaviour, the {(*R*)-NEPIA}$^-$ ligand was incorporated. The PL of **17** was recorded at various temperatures ranging from 5 to 295 K (Figure 3.3.8). In contrast to **15** and **18** complex **17** shows distinct luminescence at room temperature. Based on the integral ratio of PL Intensity vs temperature graph, at 5 K (assuming 95 %) and at 295 K the estimated quantum efficiency at 295 K can be estimated quantum yield of ϕ(295 K) = 26 %. Also, in difference to complex **15** and **18** an intense emission band centred at 500 nm is observed, which does not shift upon raising the temperature to 295 K. The broad emission band exhibits a FWHM of 75 nm at low temperature and slightly widens to 90 nm upon raising the temperature. The emission at 500 nm can be assigned to fluorescence with an decay time of τ < 10 ns Additionally, a second very broad emission is detected centred at 700 nm can be attributed to phosphorescence from the triplet manifold with a decay times of τ(5 K) ≈ 16 msec and τ(295 K) ≈ 5 msec. This interesting combination of PL components in the PL spectrum, to some extent is likely due to intramolecular (3.9189 Å) and intermolecular (4.1385(1) Å) π- stacking between the phenyl groups attached to the P2 and the naphthyl groups attached to the N4 atom of the NPN back bone, respectively.

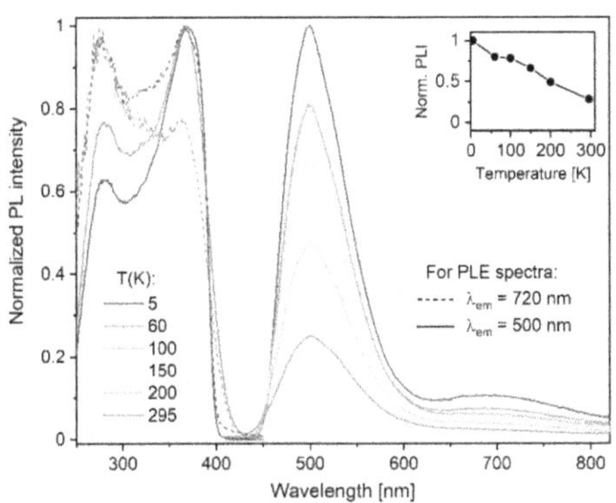

Figure 3.3.8 Photoluminescence emission (PL) and excitation (PLE) spectra (left) of **17** and integrated PL intensities plotted against temperatures from 5 K to 295 K (top right).

The study of the PL behaviour was further extended to the zinc complexes. In this regard, complexes **16** and **19** were selected and present different N-substituents on the NPN backbone.

The PL of complex **16** was recorded at temperatures ranging from 20 to 295 K. The PL spectrum exhibits a broad emission band centred at 529 nm at 20 K, which shows a red shift to 548 nm at 295 K. The broad emission bands exhibit FWHM values ranging from 100 to 133 nm. Furthermore, the integrated PL intensity decreases monotonically in contrast to the exponential decay assigned to TADF observed for the isostructural homoleptic calcium complex **12**. The PL of complex **16** is mostly assigned to a fluorescence component and the corresponding intensity decreases upon rising the temperature (Figure 3.3.9).

Besides, the structurally different heteroleptic zinc complex **19** exhibits blue green PL when exposed to UV light at room temperature. Variable temperature PL measurements revealed an intense emission band centred at 465 nm at 5 K. Upon increasing the temperature, the PL intensity decreases drastically. The difference in the PL behaviour of **19** compared to that of **16** is possibly due to the heteroleptic nature of complex **19**. At 5 K, the phosphorescence component is observed with a decay time of 25 msec whereas, at 100 K, a relatively faster decay occurs.

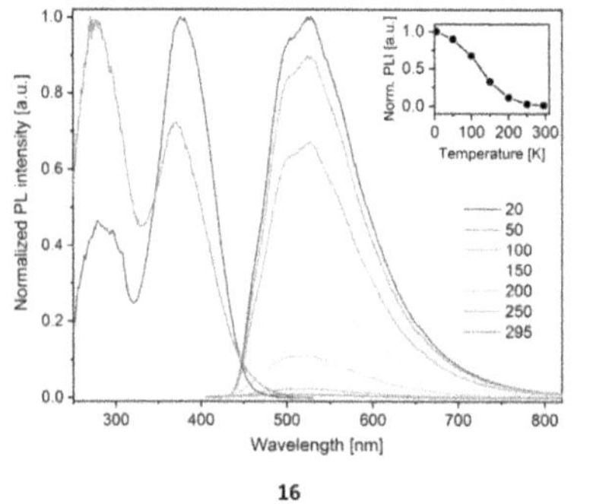

16

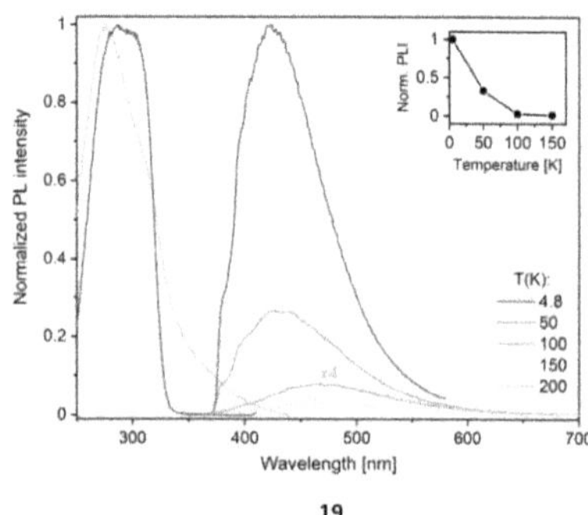

19

Figure 3.3.9 Photoluminescence emission (PL) and excitation (PLE) spectra of **16** (left) and **19** (right). For clarity in the PL spectrum of **19**, the emission bands at 100 and 150 K have been four-fold magnified. The corresponding integrated PL intensities are plotted at given temperatures (top right).

3.4 Group 13 metal complexes of enantiopure iminophosphonamide

Complexes **23-27** in the following section have already been published:

Bhupendra Goswami, Ravi Yadav, Christoph Schoo and Peter W. Roesky, Neutral and cationic enantiopure group 13 iminophosphonamide complexes, *Dalton Trans.*, **2020**, *49*, 675-681 (Schemes 3.4.4-3.4.7 and figures 3.4.4-3.4.9 are adopted from Ref. [144] with permission from the royal society of chemistry).

3.4.1 Introduction

Complexes containing group 13 metals have a long standing interest in various homogeneous catalyses involving polar substrates and Lewis acid type activation.[142] In this regard, the reactivity of the group 13 metal complexes can be modulated by fine tuning the electronic and steric properties of the respective ligand system.

Among group 13 metal complexes of iminophosphonamide ligands, mostly achiral analogue has been explored. In contrast, the coordination chemistry of chiral group 13 iminophosphonamides has not been explored prior to this work.

Therefore, the coordination behaviour of group 13 metal compounds towards enantiopure iminophosphonamides is studied, using the C_2 symmetrical (*R*)-HPEPIA and unsymmetrical (*R*)-HPE^{Dipp}PIA enantiopure ligands (Chart 3.4.1).

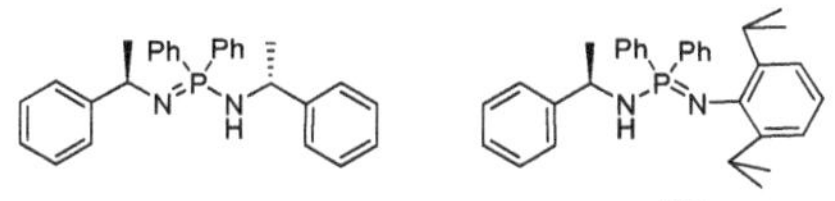

(*R*)-HPEPIA (*R*)-HPE^{Dipp}PIA

Chart 3.4.1 The enantiopure iminophosphonamide ligands used to synthesize group 13 complexes.

3.4.2 Synthesis of organo-aluminium complex of enantiopure iminophosphonamide and its reactivity

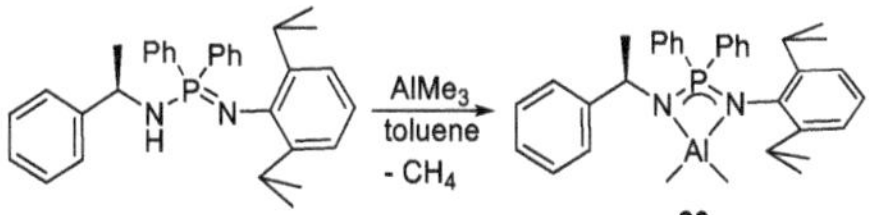

(*R*)-HPE^{Dipp}PIA → $AlMe_3$, toluene, - CH_4 → **20**

Scheme 3.4.1 Synthesis of [{(*R*)-PE^{Dipp}PIA}$_2$AlMe$_2$] (**20**).

Addition of (*R*)-HPE^{Dipp}PIA to $AlMe_3$ in toluene resulted in an exothermic reaction and concomitant elimination of methane, which leads to the formation of the heteroleptic aluminium dimethyl complex [{(*R*)-PE^{Dipp}PIA}AlMe$_2$] (**20**) (Scheme 3.4.1). The ^{1}H NMR spectrum of **20** is apparent to its nonsymmetric nature. More precisely, in the ^{1}H NMR spectrum, a doublet of quartets at δ = 3.86 ppm ($^3J_{PH}$ = 17.08 Hz, $^3J_{HH}$ = 6.67 Hz), is assigned to Ph(C*H*)(CH$_3$) which was further confirmed by ^{1}H{^{31}P} NMR spectroscopy. The Al-(C*H*$_3$)$_2$ resonances appear as two sharp singlets at δ = 0.16 and δ = 0.10 ppm due to the non-symmetrical nature of the ligand on Al centre. In the ^{13}C{^{1}H} NMR spectrum of complex **20** two singlets at δ = -4.4 ppm and δ = -4.8 ppm are assigned to the Al-(*C*H$_3$)$_2$ group, which confirms the non-symmetric nature of the complex. In the ^{31}P{^{1}H} spectrum of complex **20** one single peak at δ = 36.3 ppm is observed, which is considerably downfield shifted as compared to its precursor (δ = -10.9 ppm for (*R*)-HPE^{Dipp}PIA) and in the typical range for similar achiral complexes.[25c]

The molecular structure of complex **20** was determined by single crystal X-ray diffraction analysis. Complex **20** crystallized in the chiral orthorhombic space group $P2_12_12_1$ with one molecule in the asymmetric unit (Figure 3.4.1). The structure of **20** exhibits a distorted tetrahedral geometry in which the bidentate {(*R*)-PE^{Dipp}PIA}$^-$ ligand bound to an aluminium dimethyl moiety. The average P-N bond length (1.622(2) Å) lies in between that of single and double bonds of phosphorus and nitrogen indicating delocalisation of the π-electrons.[25c,103b]

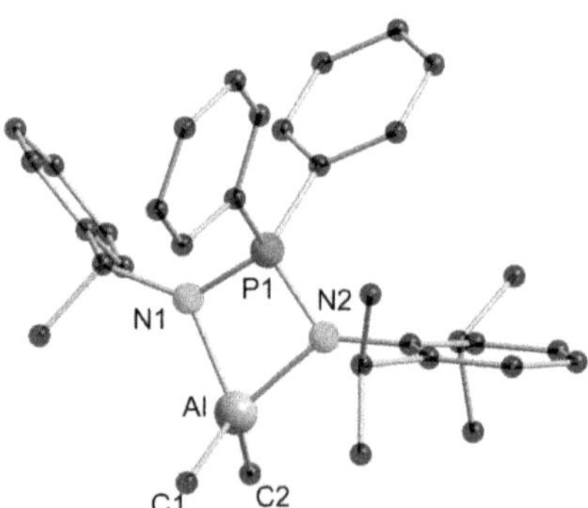

Figure 3.4.1 Molecular structure of complex **20** in the solid state. All hydrogen atoms are omitted for clarity. Selected bond lengths (Å) and bond angles [°]: Al-N1 1.90(2), Al-N2 1.90(2), Al-C1 1.969(2), Al-C2 1.964(2), P1-N1 1.618(2), P1-N2 1.627(2); N1-Al-N2 76.61(7), N1-Al-C1 116.05(9), N1-Al-C2 112.40(9), C2-Al-C1 114.66(10), N1-P1-N2 95.36(8).

The average Al-N bond distance in complex **20** (1.90 Å) is similar to that observed in [{Ph$_2$P(NSiMe$_3$)$_2$}Al(Me)Cl] (average 1.90 Å) and [{(DippN)P(Ph$_2$)(N^{t}Bu)}AlMe$_2$] (average 1.94 Å).[25c] Moreover, the central Al atom makes a planar four-membered N$_2$PAl metallocycle with the NPN backbone. The N1-Al-N2 bite angle (76.61(7)°) in complex **20** is similar to that in previous reports.[25c]

Considering the high reactivity of cationic complexes in catalysis, made to attempt the synthesis of cationic methyl-Al complex, $[\{(R)\text{-PE}^{Dipp}\text{PIA})\}\text{AlMe}]^+[\text{MeB}(C_6F_5)_3]^-$. In this regard complex **20** was reacted with $B(C_6F_5)_3$, which is a typical methyl abstracting reagent. However, this reaction resulted in the formation of an unexpected neutral complex $[\{(R)\text{-PE}^{Dipp}\text{PIA}\}\text{Al(Me)}C_6F_5]$ (**21**) (Scheme 3.4.2). The formation of complex **21** may be attributed to the methyl/C_6F_5 exchange from the presumed intermediate $[\{(R)\text{-PE}^{Dipp}\text{PIA})\}\text{AlMe}]^+[\text{MeB}(C_6F_5)_3]^-$ (Scheme 3.4.2). Unfortunately, the intermediate cationic product could not be obtained. This kind of exchange product has also been observed previously in the Lewis acid mediated synthesis of cationic aluminium species stabilised by weakly coordinating anions.[25c,143]

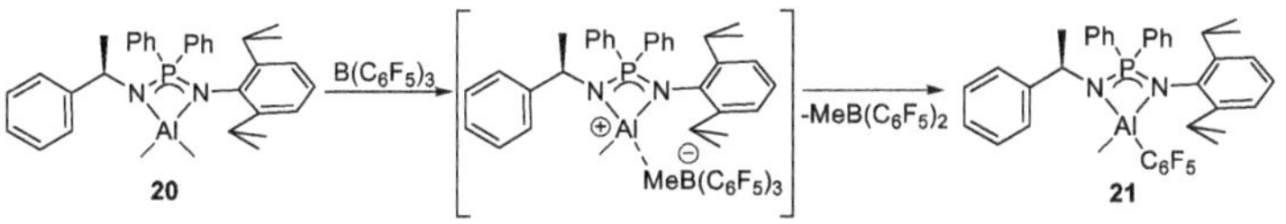

Scheme 3.4.2 Synthesis of $[\{(R)\text{-PE}^{Dipp}\text{PIA}\}\text{Al(Me)}C_6F_5]$ (**21**).

The ^{1}H NMR spectrum of **21** reveals expected resonances and suggests formation of single product. Characteristic peak at $\delta = 1.63$ ppm could be assigned to Ph(CH)CH_3, which appeared as a doublet of doublets ($^3J_{HH} = 6.72$ Hz and $^4J_{PH} = 1.23$ Hz). The $CH(CH_3)_2$ and $CH(CH_3)_2$ protons appeared as broad resonances at $\delta = 3.37$ ($\Delta\nu_{1/2} \approx 88$ Hz) ppm and $\delta = 1.48$-0.47 ppm, respectively. Furthermore, a characteristic resonance for the Al-CH_3 group is observed at $\delta = 0.05$ ppm. In the ^{13}C{^{1}H} NMR spectrum, the aluminium bound methyl group is detected at $\delta = -4.9$ ppm. In the ^{19}F NMR spectrum of complex **21**, the observed chemical shifts at $\delta = -120.3$ ppm, $\delta = -155.3$ ppm and $\delta = -162.7$ ppm could be assigned to *o*-C_6F_5, *p*-C_6F_5 and *m*-C_6F_5, respectively. Furthermore, the ^{31}P{^{1}H} NMR spectrum showed one single resonance at $\delta = 35.4$ ppm, which is slightly shifted upfield as compared to $\delta = 36.3$ ppm for complex **20**.

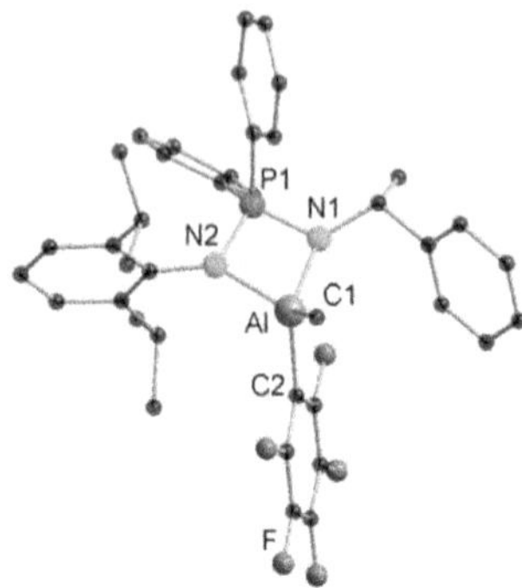

Figure 3.4.2 Molecular structure of complex **21** in the solid state. All hydrogen atoms are omitted for clarity. Selected bond lengths (Å) and bond angles [°]: Al-N1 1.920(3), Al-N2 1.911(4), Al-C1 1.956(5), P1-N1 1.633(4), P1-N2 1.623(3); N2-Al-N1 77.6(2), N2-Al-C2 111.1(2), N2-Al-C1 120.6(2), C1-Al-C2 112.2(2), N1-P1-N2 95.1(2).

Single crystals of **21** suitable for X-ray analysis were obtained from a hot *n*-heptane solution. It crystallized in the chiral monoclinic space group $P2_1$ (Figure 3.4.2). Similar to complex **20**, the aluminium centre in **21** also adopts a distorted tetrahedral geometry. The Al-N bond distances (1.911(4)-1.920(3) Å) in **21** lie in the similar range as observed for complex **20**. The Al-Me bond length of 1.956(5) Å is in consistence to that of **20** (average 1.966(2) Å). Furthermore, the aluminium centre in complex **21** is slightly above to the N_2P plane with a dihedral angle of 9.56(2)° relative to the N_2Al plane. This slight deviation possibly minimises the steric repulsion between the C_6F_5 moiety and the large Dipp substituents. The N1-Al-N2 bite angle in **21** (77.6(2)°) is similar to the one in **20** (76.61(7)°) and also in consistence with related complexes.[42a]

3.4.3 Synthesis of enantiopure iminophosphonamide group 13 halide complexes

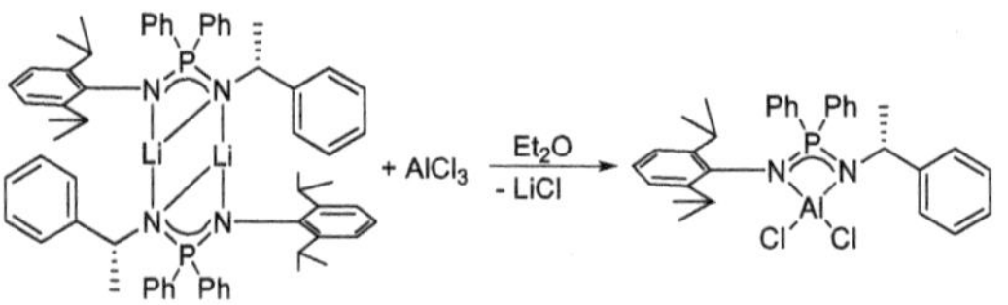

Scheme 3.4.3 Synthesis of [{(*R*)-PE^{Dipp}PIA}$AlCl_2$] (**22**).

The di-chloro substituted aluminium complex **22** was synthesized in a salt metathesis reaction between [{(*R*)-PE^{Dipp}PIA}Li]$_2$ and $AlCl_3$ using an appropriate stoichiometric ratio (Scheme 3.4.3). Successful formation of [{(*R*)-PE^{Dipp}PIA}$AlCl_2$] (**22**) was evidenced by facile precipitation of LiCl during the course of reaction. The ^{1}H NMR spectrum of complex **22** shows a characteristic doublet of doublets at δ = 1.79 ppm for the Ph(CH)CH_3 protons of the ligand backbone. Owing to the hindered rotation of the ligand backbone, the $CH(CH_3)_2$ protons show a broad resonance in the range of δ = 1.61-0.03 ppm. The corresponding methine protons, Ph(*CH*)CH_3 and $CH(CH_3)_2$, also appeared as a multiplet resonance in the range of δ = 3.89-3.24 ppm. In the $^{31}P\{^{1}H\}$ NMR spectrum of **20** one single resonance appeared at δ = 40.0 ppm, which is significantly downfield shifted (δ = 2.9 ppm) as compared to that of the starting precursor $[\{(R)\text{-PE}^{Dipp}\text{PIA}\}\text{Li}]_2$.

The solid-state structure of complex **22** was analysed by X-ray diffraction studies. Complex **22** crystallized in orthorhombic $P2_12_12_1$ space group with one molecule in the asymmetric unit (Figure 3.4.3). The molecular structure of **22** shows that the aluminium centre is in a tetracoordinated sphere surrounded by two nitrogen atoms of the iminophosphonamide ligand and two chlorine atoms. The N1-Al-N2 (79.51(10)°) bond angle in complex **22** is wider than those observed for complexes **20** (76.61(7)°) and **21** (77.6(2)°), whereas narrower than that in $[\{Ph_2P(NSiMe_3)_2\}Al(Me)Cl]$ (80.04(5)°).[42a] The central aluminium atom in complex **22** is slightly out-of-plane respective to the NPN backbone, with a dihedral angle of 5.08(2)°, in contrast to the coplanarity observed in case of complex **20**.

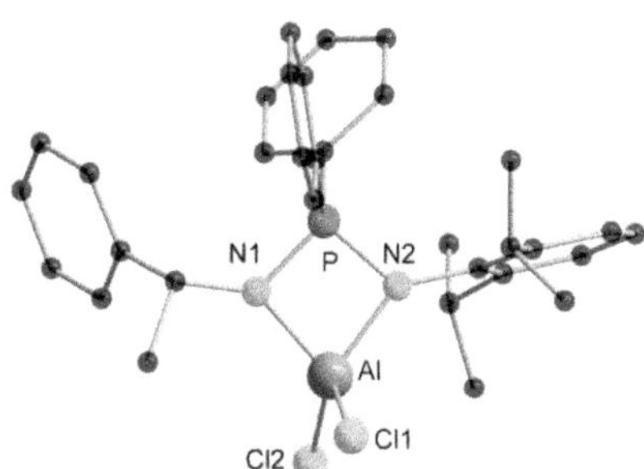

Figure 3.4.3 Molecular structure of complex **22**. All hydrogen atoms are omitted for clarity. Selected bond lengths (Å) and bond angles [°]: Al-N1 1.864(2), Al-N2 1.873(2), Al-Cl1 2.1316(11), Al-Cl2 2.1086(11), P-N1 1.623(2), P-N2 1.639(2); N1-Al-N2 79.51(10), N2-Al-Cl1 110.09(8), N2-Al-Cl2 124.60(9), N1-Al-Cl1 118.06(8), Cl2-Al-Cl1 109.50(5), N1-P-N2 94.22(12).

The average Al-N bond distance of 1.87(2) Å in complex **22** is smaller than that in **20** (1.90(2) Å) and **21** (1.92(3) Å). The P-N bond lengths in complex **20** (1.623(2) Å, 1.639(2) Å) lie in between single and double bond lengths. Such P-N bond lengths suggest the expected complete delocalisation of charge over the NPN backbone.[103b] Furthermore, upon complexation, the NPN bond angle of (*R*)-HPE^{Dipp}PIA (116.7(2)°) significantly decreases to 94.22(12)° in complex **22**.

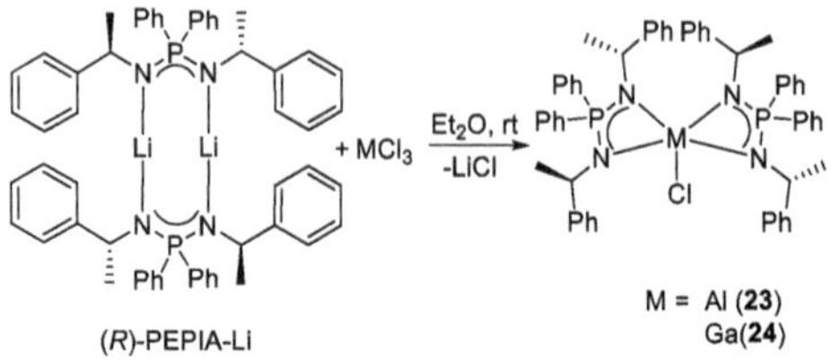

(*R*)-PEPIA-Li + MCl_3 $\xrightarrow[-LiCl]{Et_2O, rt}$ M = Al (**23**), Ga(**24**)

Scheme 3.4.4 Synthesis of [{(*R*)-PEPIA}$_2$MCl] (M = Al (**23**), Ga (**24**)).

The mono-halide substituted aluminium complex, [{(*R*)-PEPIA}$_2$AlCl] (**23**), was synthesised by carrying out a salt metathesis reaction between $AlCl_3$ and [{(*R*)-PEPIA}Li]$_2$ in a stoichiometric ratio of 1:1, respectively (Scheme 3.4.4). Reaction progress was indicated by a gradual precipitation of LiCl. After filtration and standard workup procedure, complex **23** was obtained as a white solid in quantitative yield. The formation of **23** was confirmed by various analytical techniques such as multinuclear NMR spectroscopy, IR, elemental analysis and single crystal X-ray diffraction. The recorded elemental analysis is in agreement with the molecular formula of **23**. Besides, the $^{31}P\{^1H\}$ NMR spectrum of complex **23** shows a single peak at δ = 35.6 ppm, which is significantly shifted downfield compared to the resonance at δ = 29.7 ppm observed for [{(*R*)-PEPIA}Li]$_2$. In the ^1H NMR spectrum of **23**, the methyl resonance (Ph(CH)C*H*$_3$) appears as a broad resonance at δ = 1.89 ($\Delta\nu_{1/2} \approx$ 40 Hz) ppm. The corresponding methine proton (Ph(C*H*)CH$_3$) is as well observed as a broad signal at δ = 5.17 ($\Delta\nu_{1/2} \approx$ 140 Hz) ppm. In order to understand the dynamic behaviour, a variable temperature NMR study of **23** was carried out in thf-d_8, in the temperature range from 283 K to 173 K (Figure 3.4.4).[144] At ambient temperature, the Ph(CH)C*H*$_3$ group appeared as a doublet at δ = 1.59 ppm ($^3J_{HH}$ = 6.58 Hz) and the broad resonance for Ph(C*H*)CH$_3$ is detected in the area of δ = 4.3-5.3 ppm. Upon cooling down the NMR sample from 253 to 243 K, the doublet of Ph(CH)C*H*$_3$ (δ = 1.59 ppm) at room temperature, splits into two individual resonances.

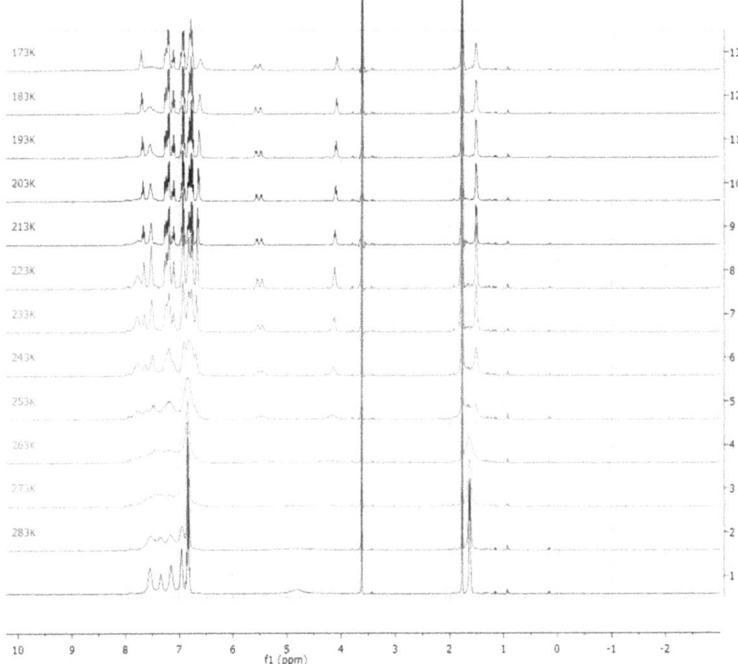

Figure 3.4.4 Stacked variable temperature ^{1}H NMR (300 MHz, thf-d_8, 298-173 K) spectra of complex **23**.

One doublet is detected at δ = 1.47 ppm ($^{3}J_{HH}$ = 6.40 Hz) at 203 K, whereas the other one is overlapping with the solvent peak at δ = 1.73 ppm. Furthermore, in the low temperature range (243 - 173 K), the broad resonance of Ph(C*H*)CH_3 splits into two rather well-resolved resonances at δ = 4.2 and δ = 5.5 ppm with equal integral ratio.

The solid state structure of complex **23** was further confirmed by single crystal X-ray diffraction studies. Complex **23** crystallized in the orthorhombic chiral space group $P2_12_12$ with half of a molecule in the asymmetric unit (Figure 3.4.5). The aluminium centre in complex **23** is pentacoordinated, surrounded by four nitrogen atoms of the {(*R*)-PEPIA}$^-$ iminophosphonamide ligand, and chlorine atom (Al-Cl = 2.174(2) Å), resulting in a distorted trigonal bipyramidal coordination geometry. The N2 and N2' lie in axial positions, whereas the N1, N1' and Cl atoms occupy the equatorial positions. The equatorial bond angles N1-Al-N1' (126.4(2)°) and N1-Al-Cl (116.81(11)°) show deviations from the expected bond angle of 120° for a trigonal bipyramidal (tbp) geometry.

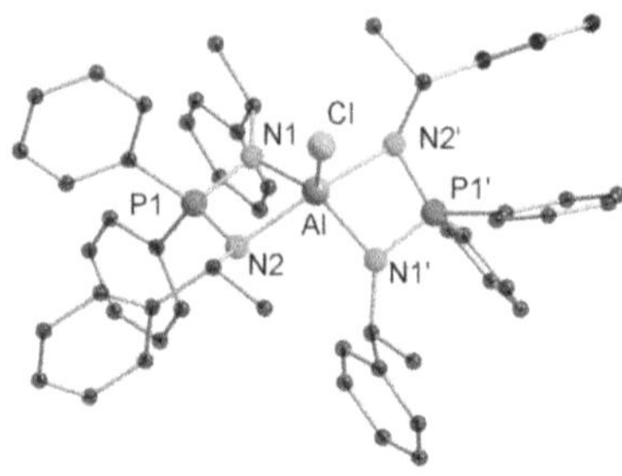

Figure 3.4.5 Molecular structure of complex **23** in the solid state. All the hydrogen atoms are omitted for clarity. Selected bond lengths (Å) and bond angles [°]: Al-N1 1.909(3), Al-N2 2.040(3), Cl-Al 2.174(2), P1-N1 1.620(3), P1-N2 1.619(3); N1-Al-N1′ 126.4(2), N2-Al-N2′ 166.6(2), N1-Al-N2 75.13(13), N1′-Al-N2 98.69(14), N1-Al-Cl 116.81(11), N2-Al-Cl 96.72(10), P1-Al-P1′ 142.76(8), N1-P1-N2 96.2(2).

Furthermore, the axial bond angle N2-Al-N2′ of 166.6(2)° also shows a deviation from the ideal value of 180°. This distortion is clearly arising due to acute N1-Al-N2 bite angle (75.13(13)°). The aluminium centre in complex **23** lies out of the planes defined by the of NPN backbones with a P1-Al-P1′ angle of 142.76(8)°. The axial bond length (Al-N2 2.040(3) Å) is longer as compared to the equatorial bond length (Al-N1 1.909(3) Å).

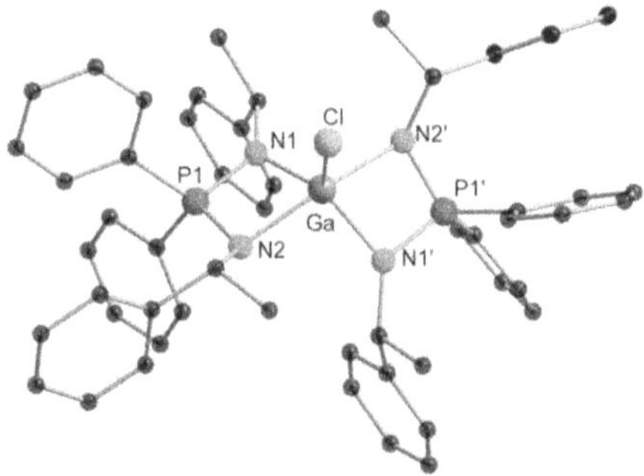

Figure 3.4.6 Molecular structure of complex **24** in the solid state. All the hydrogen atoms are omitted for clarity. Selected bond lengths (Å) and bond angles [°]: Ga-N1 1.950(6), Ga-N2 2.087(5), Ga-Cl 2.240(2), P1-N1 1.617(6), P1-N2 1.625(7); N1-Ga-Cl 116.2(2), N2-Ga-Cl 96.88(13), N2-Ga-N2′ 166.2(3), N1-Ga-N1′ 127.7(3), N1-Ga-N2 73.4(2), N1-Ga-N2′ 100.4(2), N1-P1-N2 96.4(3).

Following the similar synthetic procedure as described for **23**, the reaction of $GaCl_3$ and [{(*R*)-PEPIA}Li]$_2$ in an appropriate stoichiometric ratio led to the formation of [{(*R*)-PEPIA}$_2$GaCl] (**24**) (Scheme 3.4.4).

The ^{1}H NMR spectrum of **24**, exhibits the expected signals. The $^{31}P\{^{1}H\}$ NMR spectrum of **24** exhibit a single resonance at δ = 36.5 ppm which is shifted downfield as compared to the signals at δ = 35.6 ppm (**23**) and δ = 29.7 ppm in [{(*R*)-PEPIA}Li]$_2$. The molecular structure of complex **24** in the solid state was determined by single crystal X-ray diffraction (Figure 3.4.6). Complex **24** exists in a distorted trigonal bipyramidal geometry similar to **23**. The Ga-Cl bond length (2.240(2) Å) in **24** is slightly longer than the Al-Cl (2.174(2) Å) bond length observed in **23**, which can be explained by the larger ionic radius of Ga. The N-M-N bite angle in **24** (N1-Ga-N2 73.4(2)°) is slightly narrower than the N1-Al-N2 angle of 75.13(13)° in **23**.

3.4.4 Synthesis of enantiopure iminophosphonamide group 13 cationic complexes

As the reaction of **20** and $B(C_6F_5)_3$ towards complex **21** did not result in the desired cationic species, but led to a neutral complex (*vide supra*), a new pathway towards charge-separated group 13 metal complexes bearing the (*R*)-PEPIA ligand was developed. Therefore, reaction of the monochloride complexes **23** and **24** with $GaCl_3$ and $AlCl_3$ led to the formation of the expected cationic complexes, [{(*R*)-PEPIA}$_2$Al]$^+$[$GaCl_4$]$^-$ (**25**), and [{(*R*)-PEPIA}$_2$Ga]$^+$[$AlCl_4$]$^-$ (**26**), respectively (Scheme 3.4.5). These cationic complexes were further characterised by using various analytical techniques such as multinuclear NMR, IR and elemental analysis. In case of complex **25**, the Ph(C*H*)CH_3 protons of the ligand backbone appear as a multiplet in the range δ = 4.35-4.27 ppm in the ^{1}H NMR spectrum, whereas the corresponding resonance for **26** appears as a doublet of quartets at δ = 4.31 ($^{3}J_{PH}$ = 9.42 Hz, $^{3}J_{HH}$ = 6.63 Hz) ppm.

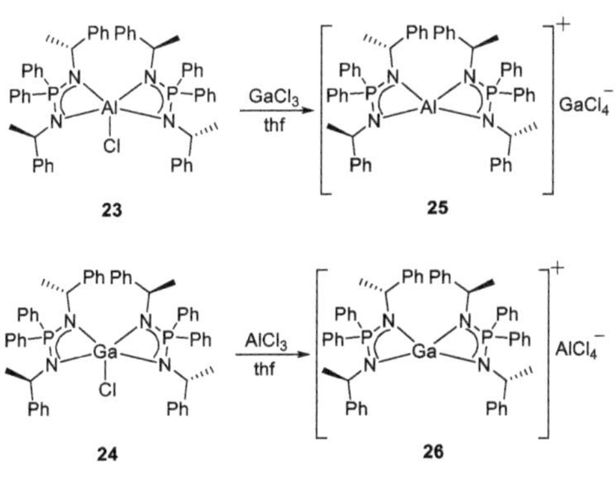

Scheme 3.4.5 Synthesis of [{(*R*)-PEPIA}$_2$Al]$^+$[$GaCl_4$]$^-$ (**25**), and [{(*R*)-PEPIA}$_2$Ga]$^+$[$AlCl_4$]$^-$ (**26**).

Furthermore, the Ph(CH)CH_3 resonance appeared as a doublet of doublets at δ = 1.56 ppm ($^3J_{HH}$ = 6.67 Hz, $^4J_{PH}$ = 1.44 Hz) and δ = 1.55 ppm ($^3J_{HH}$ = 6.71 Hz, $^4J_{PH}$ = 1.85 Hz) for **25** and **26**, respectively. The appearance of one single signal at δ = 43.3 ppm (**25**) and δ 48.6 ppm (**26**) in the $^{31}P\{^1H\}$ NMR spectrum is indicative of the purity of the species. These resonances are shifted downfield compared to those of the corresponding monohalide complexes (δ = 35.6 ppm (**23**) and δ = 36.5 ppm (**24**)).

The molecular structures in the solid state were further confirmed by single crystal X-ray diffraction. Both complexes **25** and **26** crystallize in the trigonal chiral space group $P3_121$ with two halves of the molecule in the asymmetric unit. Complexes **25** and **26** are isostructural and the central metals in both the cationic and the counter anionic parts adopt a distorted tetrahedral geometry (Figure 3.4.7 and Figure 3.4.8).

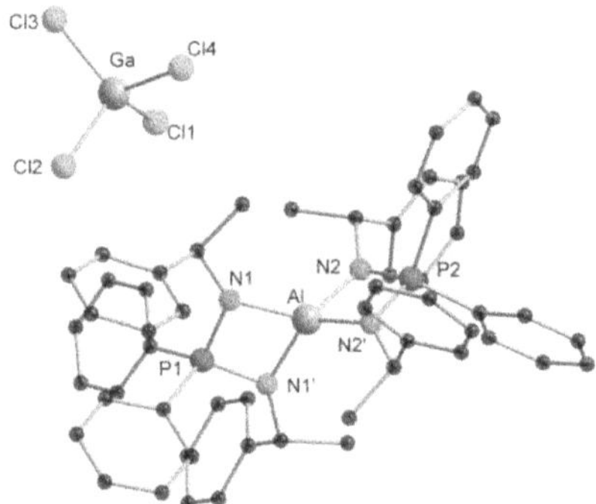

Figure 3.4.7 Molecular structure of complex **25** in the solid state. All hydrogen atoms are omitted for clarity. Selected bond lengths (Å) and bond angles [°]: Al-N1 1.857(4), Al-N2 1.871(4), Ga-Cl1 2.166(2), Ga-Cl2 2.171(2), Ga-Cl3 2.158(2), Ga-Cl4 2.15(2), P1-N1 1.631(4), P2-N2 1.629(4); N1-Al-N1' 79.2(2), N1-Al-N2 126.3(2), N1-Al-N2' 126.3(2), N2-Al-N2' 79.7(2), P2-Al-P1 180.0, Cl1-Ga-Cl2 110.89(6), Cl3-Ga-Cl1 109.87(7), Cl3-Ga-Cl2 106.93(7), Cl4-Ga-Cl1 108.97(7), Cl4-Ga-Cl2 108.19(9), Cl4-Ga-Cl3 112.0(9), N1-P1-N1' 93.0(3), N2-P2-N2' 94.7(3).

The difference in the ionic radii of Al and Ga is reflected by the different M-Cl and M-N bond lengths in the respective anion and cation. In detail, the Ga-N distance of 1.931(6) Å in **26** is significantly elongated compared to that in **25** (1.864(4) Å). Similarly, in the counter anion $[MCl_4]^-$, the average M-Cl (M = Ga(**25**), Al(**26**)) bond distance of 2.123(4) Å for complex **26** is smaller compared to 2.161(2) Å for complex **25**. These values are however in the expected range with literature reports.[145]

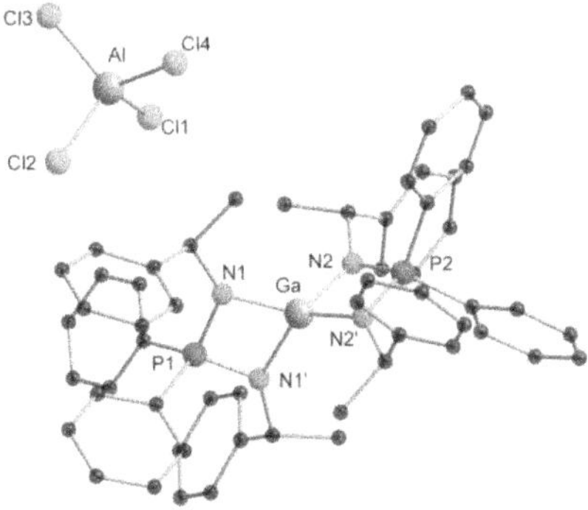

Figure 3.4.8 Molecular structure of **26** in the solid state. All the hydrogen atoms are omitted for clarity. Selected bond lengths (Å) and bond angles [°]: Ga-N1 1.923(6), Ga-N2 1.938(6), Al-Cl1 2.129(3), Al-Cl2 2.119(4), Al-Cl3 2.134(4), Al-Cl4 2.109(4), P1-N1 1.629(6), P2-N2 1.626(6); N1-Ga-N1' 76.9(4), N1-Ga-N2 126.5(3), N2-Ga-N2' 77.2(4), N1-Ga-N2' 129.0(3), Cl1-Al-Cl3 110.7(2), Cl2-Al-Cl1 109.8(2), Cl2-Al-Cl3 107.4(2), Cl4-Al-Cl1 109.2(2), Cl4-Al-Cl2 112.2(2), Cl4-Al-Cl3 107.5(2), N1-P1-N1' 94.4(5), N2-P2-N2' 96.1(5).

The central metal atom forms two four-membered metallacycles with the NPN backbone, in which the N_2PM planes are twisted with dihedral angles of 89.97(1)° (**25**) and 87.53(2)° (**26**) with respect to the NPN backbone. The distortion in both of these complexes is attributed to corresponding narrow bite angles (N1-Al-N1' 79.2(2), N2-Al-N2' 79.7(2)) for **25** and (N1-Ga-N1' 76.9(4) and N2-Ga-N2' 77.2(4)) for **26**.

3.4.5 Synthesis of an enantiopure iminophosphonamide aluminium hydride complex

The aluminium monohydride complex [{(*R*)-PEPIA}$_2$AlH] (**27**) was prepared by reaction of (*R*)-HPEPIA with $LiAlH_4$ in a molar ratio of 2:1 (Scheme 3.4.6). Formation of **27** was further confirmed by multinuclear NMR spectroscopy, IR, elemental analysis. In the ^{1}H NMR spectrum of **27**, the resonance associated with the Al-*H* moiety could not be detected, which is most likely attributed to the quadrupole moment of the ^{27}Al nuclei, leading to a fast relaxation of the hydride signal.

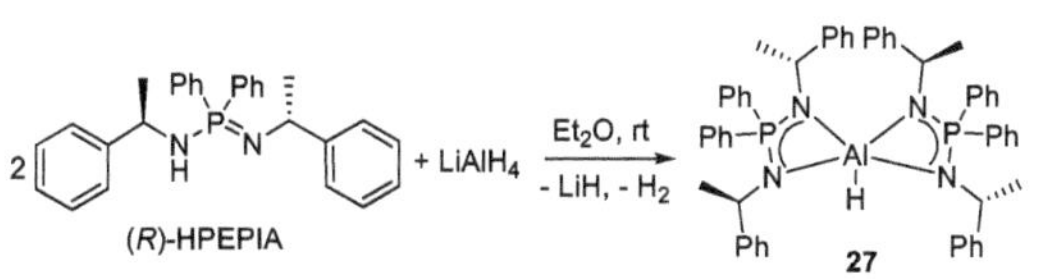

2 (*R*)-HPEPIA + $LiAlH_4$ → (Et_2O, rt; - LiH, - H_2) **27**

Scheme 3.4.6 Synthesis of [{(*R*)-PEPIA}$_2$AlH] (**27**).

In order to confirm the presence of a hydride in complex **27**, the isotopic analogue **27-d** was synthesized by reacting (*R*)-HPEPIA with $LiAlD_4$ in the appropriate stoichiometric ratio. The ^1H NMR spectrum of the deuterium labelled **27-d** showed a ^1H NMR pattern comparable to that of the analogous complex **27**. Besides, in the IR spectrum of complex **27**, a band at 1693 cm^{-1} was assigned to the Al-H stretching frequency. The corresponding IR spectrum of the deuterium analogue (**27-d**) does not exhibit any band in this wavenumber range. These findings were further supported by simulations.[144] In the ^1H NMR spectrum of complex **27**, the broad resonance at δ = 4.99 ($\Delta\nu_{1/2}$ ≈ 55 Hz) ppm is assigned to the methine protons (Ph(C*H*)CH_3). However, the corresponding methyl (Ph(CH)CH_3) protons appears as a clear doublet at δ = 1.90 ppm with a coupling constant $^3J_{HH}$ = 6.45 Hz. The appearance of a single set of resonances for the methine and methyl protons in the ^1H NMR spectrum of complex **27** indicates its symmetric nature in solution in contrast, the ^1H NMR spectrum of (*R*)-HPEPIA, shows two different sets of resonances corresponding to the methine (Ph(C*H*)CH_3) and methyl (Ph(CH)CH_3) group due to *E/Z* tautomerization. The $^{31}P\{^1H\}$ NMR spectrum of **27** shows a single peak at δ = 33.6 ppm, which is significantly shifted downfield in comparison to the parent (*R*)-HPEPIA (δ = 2.7 ppm).

Single crystals of **27** were grown from Et_2O at room temperature. It crystallizes in the orthorhombic chiral space group $P2_12_12$ with half of the molecule in the asymmetric unit (Figure 3.4.9). The pentacoordinated aluminium centre adopts a distorted trigonal bipyramidal coordination geometry. This structure is similar to that of previously reported aluminium monohydride complexes based on amidinate,[146] guanidinate,[147] and iminophosphonamide ligands.[42a] The solid state structure showed that the aluminium centre in complex **27** is surrounded by four nitrogen atoms of the bidentate iminophosphonamide ligands (making two four-membered metallacycles) and one hydrogen atom (which was localized in the difference Fourier map and was freely refined). The P1···Al···P1′ vector is clearly bent (139.9(4)°), and the dihedral angle between the N-M-N and N-P-N planes is 7.285°.

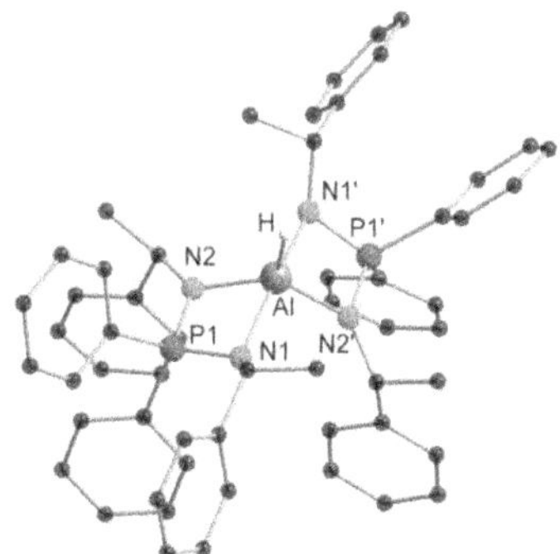

Figure 3.4.9 Molecular structure of complex **27** in the solid state. All hydrogen atoms except the hydride are omitted for clarity. Selected bond lengths (Å) and bond angles [°]: Al-N1 2.040(2), Al-N2 1.932(2), Al-H 1.48(4), P1-N1 1.621(2), P1-N2 1.622(2); N1-Al-N1' 165.03(9), N2-Al-N1' 97.68(7), N2-Al-N1 74.93(7), N2-Al-N2' 121.99(11), N1-Al-H 97.49(5), N2-Al-H 119.01(5), P1-Al-P1' 139.9(4), N2-P1-N1 96.46(9).

Similar to complex **23** and **24**, the distortion in complex **27** is attributed to the N1-Al-N2 bite angle of 74.93(7)°. The N2-Al-N1 bite angle of 74.93(7)° in complex **27** is relatively similar to the N-Al-N bite angle reported in [{$Ph_2P(NSiMe_3)_2$}$_2$AlH] (75.66(5)°).[42a]

3.5 Group 4 metal complexes of enantiopure iminophosphonamides

3.5.1 Introduction

Since the development of Kaminsky catalysts for the Ziegler-Natta polymerization, group 4 metal complexes have widely been investigated with ligands other than cyclopentadienyls and derivatives in order to alter the activity, selectivity and stability of the metal complexes.[148] In this regard, group 4 chemistry with amidinates has also been explored since the late 1980s.[149] However, corresponding chiral metal complexes featuring amidinates are rare.[77,79,150] Considering the related iminophosphonamide (NPN) ligand system, as described in section 1.2.4, only few achiral group 4 metal complexes are known. In contrast to chiral amidinates, the coordination chemistry of group 4 metal complexes bearing chiral iminophosphonamides has not been explored to the best of our knowledge. Considering this fact, novel group 4 metal complexes bearing chiral iminophosphonamides have been synthesized.

3.5.2 Synthesis of enantiopure group 4 iminophosphonamide complexes

Following the similar amine elimination route as described for the synthesis of [{(*R*)-PEPIA}Zr(NMe_2)$_3$],[138] the reaction of (*R*)-HPEPIA with tetrakis(dimethylamido)hafnium(IV) in an equimolar ratio resulted in the formation of [{(*R*)-PEPIA}Hf(NMe_2)$_3$] (**28**) (Scheme 3.5.1).

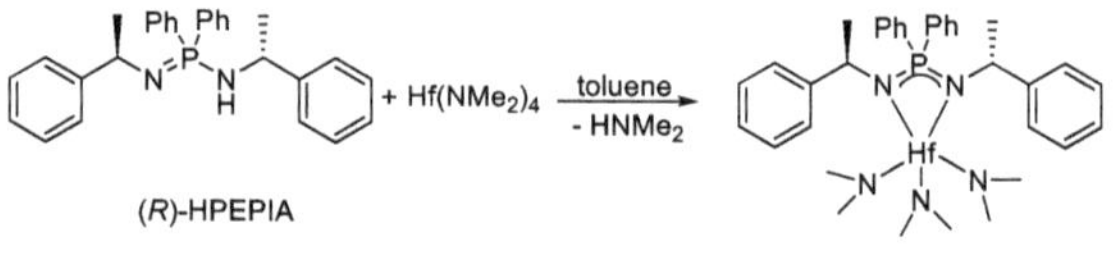

(*R*)-HPEPIA + Hf(NMe_2)$_4$ → (toluene, - $HNMe_2$) **28**

Scheme 3.5.1 Synthesis of [{(*R*)-PEPIA}Hf(NMe_2)$_3$] (**28**).

In the ^{1}H NMR spectrum of **28**, signals are observed as expected: the N(CH_3)$_2$ groups appear as a singlet at δ = 3.3 ppm, and a well-resolved characteristic doublet for Ph(CH)CH_3 is detected at δ = 1.46 ppm ($^3J_{HH}$ = 7.39 Hz) while the Ph(*CH*)CH_3 protons appear as a doublet of quartets at δ = 4.43 ppm ($^3J_{PH}$ = 17.62 Hz, $^3J_{HH}$ = 6.73 Hz). Owing to *E*/*Z* tautomerization, the corresponding methine (PhC*H*CH_3) and methyl (PhCHCH_3) protons of the ligand (*R*)-PEPIA show two sets of signals at δ = 4.73, 4.38 ppm and δ = 1.66, 1.13 ppm, respectively. Since only one set of signals is observed for complex **28**, a symmetric coordination environment around the metal is suggested in solution. The ^{31}P{^{1}H} NMR spectrum of complex **28** exhibits one single peak at δ = 36.7 ppm, which is significantly shifted downfield as compared to the signal at δ = 2.7 ppm for (*R*)-HPEPIA.[103b] The elemental analysis of **28** is also consistent with the proposed structure.

The solid-state structure of complex **28** was determined by X-ray diffraction studies. Single crystals suitable for X-ray analysis were grown from a *n*-pentane solution. Complex **28** crystallizes in the orthorhombic $P2_12_12_1$ chiral space group with one molecule in the asymmetric unit (Figure 3.5.1).

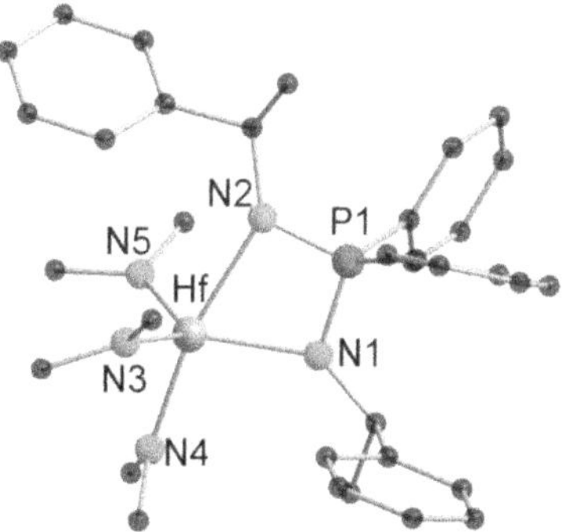

Figure 3.5.1 Molecular structure of **28** in the solid state. All hydrogen atoms are omitted for clarity. Selected bond lengths (Å) and bond angles [°]: Hf-N1 2.182(3), Hf-N2 2.315(3), Hf-N3 2.038(3), Hf-N4 2.083(3), Hf-N5 2.062(3), P1-N1 1.629(3), P1-N2 1.606(3); N1-Hf-N2 66.32(11), N1-Hf-N3 119.92(13), N2-Hf-N3 91.27(13), N3-Hf-N4 93.55(14), N3-Hf-N5 118.86(14), N1-Hf-N4 100.65(13), N2-Hf-N4 166.75(13), N1-Hf-N5 118.64(13), N2-Hf-N5 96.68(12), N4-Hf-N5 91.79(14), N1-P1-N2 99.2(2).

The central metal in **28** adopts a distorted trigonal bipyramidal geometry (tbp). The axial (N4-Hf-N2) bond angle of 166.75(13)° is significantly wider than the respective equatorial bond angles (N3-Hf-N5 118.86(14)°, N5-Hf-N1 118.64(13)°, N3-Hf-N1 119.92(13)°). This slight deviation from the ideal tbp geometry arises from the N1-Hf-N2 66.32(11)° acute bite angle. Upon coordination to the central metal, the bond angle of (*R*)-HPEPIA significantly decreases from 121.6(2)° to 99.2(2)° in complex **28**. Furthermore, the Hf-N bond lengths in the four-membered metallacycle (Hf-N1 2.182(3) Å, Hf-N2 2.315(3) Å) are longer than the Hf-N bond lengths of the amido groups (Hf-N3 2.038(3) Å, Hf-N4 2.083(3) Å, Hf-N5 2.062(3) Å) and are comparable to those of similar complexes.[138]

In order to further investigate the coordination behaviour of the (*R*)-HPE^{Dipp}PIA ligand towards group 4 (*i.e.* Zr, Hf) metals, an equimolar reaction of (*R*)-HPE^{Dipp}PIA and $M(NMe_2)_4$ (M = Zr, Hf) was conducted. The expected products [{(*R*)-PE^{Dipp}PIA}M(NMe_2)$_3$] (M = Zr (**29**), Hf (**30**)) were formed *via* an amine elimination reaction (Scheme 3.5.2).

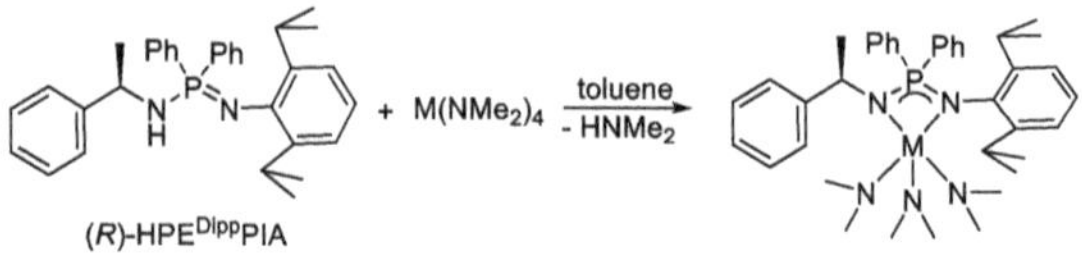

(*R*)-HPE^{Dipp}PIA + M(NMe_2)$_4$ —toluene, - $HNMe_2$→ M = Zr (**29**), Hf (**30**)

Scheme 3.5.2 Synthesis of [{(*R*)-PE^{Dipp}PIA}M(NMe_2)$_3$] (M = Zr (**29**), Hf (**30**)).

The identity of both complexes was confirmed by various analytical techniques, such as multinuclear NMR, elemental analysis, IR and single crystal X-ray crystallography. The ^{1}H NMR spectrum of both **29** and **30** are consistent with the proposed structures. For both complexes, the Ph(C*H*)CH_3 resonance appears as doublet of quartets at δ = 4.07 ppm ($^3J_{PH}$ = 17.32 Hz, $^3J_{HH}$ = 6.72 Hz) (**29**) and δ = 4.17 ppm ($^3J_{PH}$ = 16.03 Hz, $^3J_{HH}$ = 6.73 Hz) (**30**). Corresponding resonances for the methyl moiety (Ph(CH)C*H*$_3$) appear as a doublet of doublets at δ = 1.47 ppm ($^3J_{HH}$ = 6.72 Hz, $^4J_{PH}$ = 0.71 Hz) (**29**) and at δ = 1.52 ppm ($^3J_{HH}$ = 6.75 Hz, $^4J_{PH}$ = 1.10 Hz) (**30**). The appearance of one singlet in the ^{31}P{^{1}H} NMR spectrum at δ = 33.9 ppm (**29**) and δ = 33.7 ppm (**30**), significantly shifted downfield compared to the signal at δ = -10.9 ppm for (*R*)-HPE^{Dipp}PIA, indicate the purity of these complexes. Finally, the disappearance of the NH stretching frequency in the corresponding IR spectra confirms complete conversion.

For both complexes, single crystals suitable for X-ray analysis were obtained from *n*-pentane solutions. Complexes **29** and **30** are isostructural and crystallize in the orthorhombic $P2_12_12_1$ space group with two (**29**) and one (**30**) molecule in the asymmetric unit (Figure 3.5.2). The central metal atoms are five-fold coordinated, surrounded by one bidentate (*R*)-PE^{Dipp}PIA ligand and three dimethyl amide groups, and realizing a distorted trigonal bipyramidal geometry. The N2, N4 and N5 atoms lie in the equatorial plane, whereas the N1 and N3 atoms occupy the axial positions in both complexes. The distortion in both complexes is attributed to the N1-M-N2 acute bite angles of 64.43(12)° for **29** and 65.1(2)° observed for complex **30**, which are in the range of those reported for similar complexes.[67b] Furthermore, the respective central atom of the NPN backbone is slightly out of the N1-M-N2 coordination plane with a dihedral angle (N1-P1-N2 *vs* N1-Zr-N2 plane) of 6.62(2)° (**29**). In contrast, the central metal atom in **30** is almost coplanar with the NPN plane, having a dihedral angle of 0.8(2)°.

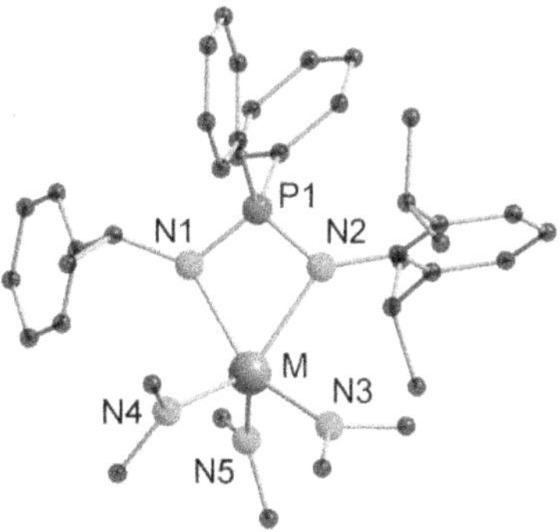

Figure 3.5.2 Molecular structures of **29** (M = Zr) and **30** (M = Hf) in the solid state. All hydrogen atoms are omitted for clarity. Selected bond lengths (Å) and bond angles [°]: For **29**: Zr1-N1 2.274(3), Zr1-N2 2.367(3), Zr1-N3 2.081(4), Zr1-N4 2.053(3), Zr1-N5 2.036(4), P1-N1 1.604(3), P1-N2 1.611(4); N1-Zr1-N2 64.43(12), N1-Zr1-N3 144.49(13), N1-Zr1-N4 95.22(13), N1-Zr1-N5 105.07(14), N2-Zr1-N3 88.39(13), N2-Zr1-N4 141.06(13), N2-Zr1-N5 116.63(13), N3-Zr1-N4 92.48(14), N3-Zr1-N5 107.6(2), N4-Zr1-N5 100.17(14), N1-P1-N2 100.7(2). For **30**: Hf-N1 2.299(6), Hf-N2 2.273(6), Hf-N3 2.073(8), Hf-N4 2.054(8), Hf-N5 2.044(8); P1-N1 1.613(7), P1-N2 1.606(7); N1-Hf-N2 65.1(2), N1-Hf-N3 151.6(3), N1-Hf-N4 95.1(3), N1-Hf-N5 103.5(3), N2-Hf-N3 91.3(3), N1-Hf-N4 95.1(3), N2-Hf-N5 114.6(3), N3-Hf-N4 93.3(3), N1-Hf-N5 103.5(3), N4-Hf-N5 104.2(4), N1-P1-N2 99.7(3).

Due to repulsion between the Dipp group and the three amido (NMe_2) groups, the ligand coordinates the metal in an asymmetric fashion, resulting in different M-N contact lengths (Zr1-N1 2.274(3) Å, Zr1-N2 2.367(3) Å) (**29**) and (Hf-N1 2.299(6) Å, Hf-N2 2.273(6) Å) (**30**). However, these values are consistent with those of related complexes.[67b] In both cases, the M-N bond lengths in the four-membered metallacycles are longer than the M-N bond distances involving the negatively charged amido groups.

3.6 Enantiopure homoleptic lanthanide complexes of iminophosphonamides

3.6.1 Introduction

Rare earth metal compounds have found useful and diversified applications in various fields such as data storage, catalytic converters and electricity generating windmills *etc.*[9] This versatility can be attributed to their interesting magnetic, chemical and optical properties.[12] Lanthanide complexes based on P,N ligands in the coordination sphere are of great interest due to their interesting catalytic activities.[151] In this regard, various ligands such as phosphinimides, R_2PNR',[152] phosphoraneiminates R_3PN,[153] phosphiniminomethanides $(RNPR'_2)_2CH$,[154] and $R_2P(NR')_2$ iminophosphonamides,[68,155] have been screened. Additionally, lanthanide complexes bearing chiral phosphanylamides, $\{N(R\text{-}^*CHMePh)(PPh_2)\}^-$, have also been reported.[156]

Bearing in mind the intriguing results obtained with chiral (*S*)-PEBA complexes of rare earth metals and their application in catalysis,[76,82a,82c,82d,130,157] we were interested to study the coordination behaviour of structurally diverse chiral iminophosphonamide ligands. Moreover, reports on iminophosphonamide lanthanide complexes are limited to only achiral iminophosphonamide ligands.[74] Herein, the coordination behaviour of $\{(R)\text{-PEPIA}\}^-$ is discussed with various lanthanide metals.

3.6.2 Synthesis of chiral homoleptic lanthanide complexes

After the successful synthesis of group 13 metal complexes with the $\{(R)\text{-PEPIA}\}^-$ ligand, the latter was further introduced to rare earth metal centres, since the lanthanide cations Ln^{3+} exhibit a similar high Lewis acidity comparable to that of group 13 elements.

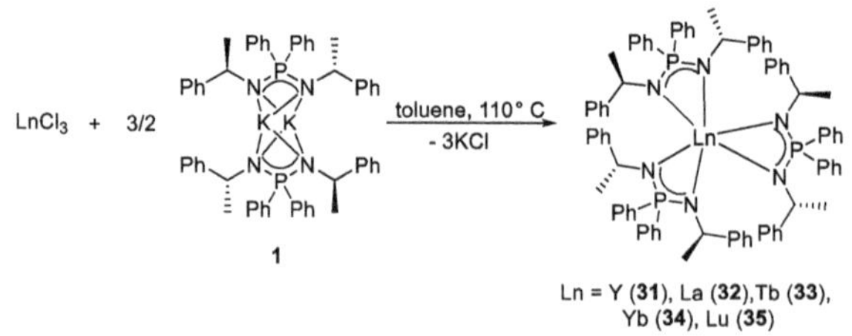

$LnCl_3$ + 3/2 **1** → (toluene, 110° C; - 3KCl)

Ln = Y (**31**), La (**32**), Tb (**33**), Yb (**34**), Lu (**35**)

Scheme 3.6.1 Synthesis of the homoleptic group 3 and lanthanide complexes **31-35**.

A salt metathesis reaction between $LnCl_3$ and [{(*R*)-PEPIA}K] in a 1:3 molar ratio led to the formation of a series of homoleptic group 3 and lanthanide complexes [{(*R*)-PEPIA}$_3$Ln] (Ln = Y (**31**), La (**32**), Tb (**33**), Yb (**34**), and Lu (**35**)) (Scheme 3.6.1).

The ^{1}H and ^{13}C{^{1}H} NMR spectra of the diamagnetic complexes suggest a symmetric arrangement of ligands upon coordination with the lanthanide metals. Due to the highly paramagnetic nature of complexes **33** and **34**, the NMR study of these complexes was not possible. The most characteristic signals were observed for Ph(C*H*)CH_3 (appearing at δ = 4.98 ppm (**31**), δ = 4.85 ppm (**32**) and δ = 5.02 ppm (**35**)) and Ph(CH)CH_3 (at δ = 2.15 ppm (**31**), δ = 2.11 ppm (**32**) and δ = 2.14 ppm (**35**)). These characteristic signals of the lanthanide complexes are downfield shifted in comparison to the starting material **1**. Furthermore, a single resonance appeared in the ^{31}P{^{1}H} NMR spectra at δ = 30.3 ppm (**31**), δ = 30.1 ppm (**32**), δ = 30.3 ppm (**35**), whereas the paramagnetic Yb complex exhibits a resonance at δ = 87.5 ppm (**34**). The presence of a single resonance in the ^{31}P{^{1}H} NMR spectra can be used as a tool to assess the purity of the complexes and suggests the formation of highly symmetric species.

The solid-state structures of these complexes were determined by single crystal X-ray diffraction. All of these complexes have similar structures, however, crystallise in different chiral space groups (Table 3.6.1).

Table 3.6.1 Selected structural parameters of complexes **31-35**.

Complex	Central metal (M^{III})	Ionic radius (Å)	Space Group	M-N bond length (Å) (avg.)	N-M-N bite angle (°) (avg.)
31	Y	1.075	$P4_12_12$	2.400(5)	97.75(2)
32	La	1.216	$P2_12_12_1$	2.571(6)	58.54(2)
33	Tb	1.095	$P2_1$	2.440(6)	61.46(2)
34	Yb	1.042	$P2_1$	2.39(2)	62.7(6)
35	Lu	1.032	$P4_12_12$	2.375(3)	63.02(7)

Each of the lanthanide ion is hexacoordinated by six nitrogen atoms of the {(*R*)-PEPIA}$^-$ ligand. In all the four-membered N-P-N-Ln metallacycles formed by coordination of the {(*R*)-PEPIA}$^-$ ligand with the metal ion, the ligands are slightly rotated against each other and form a propeller-type structure. A similar type of coordination has also been observed in [Sm{(*S*)-

PEBA}$_3$] and homoleptic lanthanide complexes of (*S*)-NEPIA.[82d,158] The central metals have no vacant sites and are completely shielded by the ligands. Since all complexes crystallize in isostructural arrangements, only the crystal structure of **31** is discussed in detail here. Complex **31** crystallizes in the tetragonal chiral space group $P4_12_12$ space group with half of the molecule in the asymmetric unit (Figure 3.6.1). The Y-N bond lengths (Y-N1 2.400(6), Y-N2 2.408(5), Y-N3 2.393(6) Å) are slightly longer than those reported for [((*S*)-NEPIA)$_3$Y] (Y-N1 2.376(4), Y-N2 2.362(4)), whereas similar to those reported for [$Ph_2P(^tBuN)_2Y(CH_2SiMe_3)$] (avg 2.407 Å).[74,82b] Furthermore, the P-N bond lengths (P1-N1 1.599(6), P1-N2 1.607(6) and P2-N3 1.605(6) Å) are also in between the range of single and double bonds, suggesting the complete delocalisation of the anionic charge over the NPN ligand backbone. The bite angles of the {(*R*)-PEPIA} ligands around the metal centre are N1-Y-N1′ 96.7(3)° and N3-Y-N2 98.8(2)°, which is significantly wider than the angle of 63.3(1)° observed in case of [$Ph_2P(^tBuN)_2Y(CH_2SiMe_3)$].

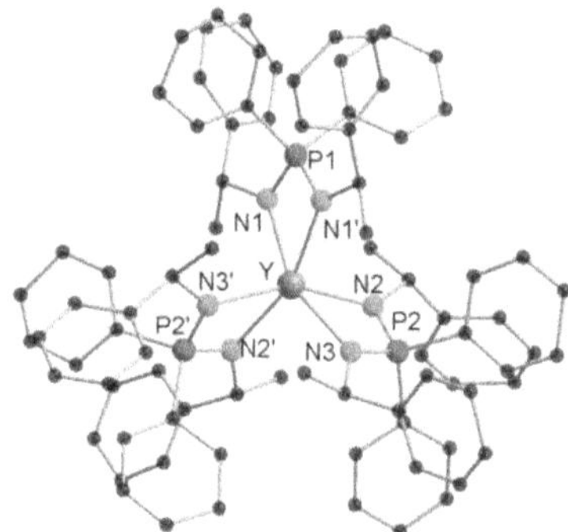

Figure 3.6.1 Molecular structure of **31** in the solid state. All hydrogen atoms are omitted for clarity. Selected bond lengths (Å) and bond angles [°] for the isostructural complexes **31-35** are given.

For **31:** Y-N1 2.400(6), Y-N2 2.408(5), Y-N3 2.393(6), P1-N2 1.607(6), P1-N1 1.599(6), P2 N3 1.605(6); N1-Y-N1′ 96.7(3), N1′-Y-N2 101.8(2), N1-Y-N2 61.6(2), N2-Y-N3 98.8(2), N3-Y-N2′ 101.3(2), N1-Y-N3 102.7(2), N1-P1-N1′ 102.5(6), N2-P2-N3 101.5(4).

For **32**: La-N1 2.572(6), La-N2 2.548(6), La-N3 2.563(6), La-N4 2.629(6), La-N5 2.553(6), La-N6 2.561(5), P1-N1 1.612(6), P1-N2 1.614(6), P2-N3 1.622(6), P2-N4 1.595(6); N1-La-N2 58.8(2), N2-La-N3 107.1(2), N3-La-N4 58.5(2), N5-La-N6 58.3(2).

For **33**: Tb-N1 2.434(6), Tb-N2 2.422(7), Tb-N3 2.454(6), Tb-N4 2.454(6), Tb-N5 2.445(7), Tb-N6 2.435(6), P1-N1 1.627(7), P1-N2 1.610(8), P2-N3 1.614(6), P2-N4 1.605(7), P3-N5 1.611(6), P3-N6 1.624(7); N1-Tb-N2 61.6(2), N3-Tb-N4 61.0(2), N5-Tb-N6 61.8(2), N1-P1-N2 100.4(3), N3-P2-N4 101.4(3), N5-P3-N6 101.6(3).

For **34**: Yb-N1 2.38(2), Yb-N2 2.37(2), Yb-N3 2.39(2), Yb-N4 2.38(2), Yb-N5 2.44(2), Yb-N6 2.40(2), P1-N1 1.61(2), P1-N2 1.62(2), P2-N3 1.63(2), P1-N2 1.610(8), P2-N4 1.61(2), P3-N5 1.63(2), P3-N6 1.61(2); N1-Yb-N2 62.5(7), N3-Yb-N4 62.9(6), N5-Yb-N6 62.7(7), N3-P2-N4 100.3(9), N1-P1-N2 99.2(9), N5-P3-N6 102.5(9).

For **35**: Lu-N1 2.386(3), Lu-N2 2.368(3), Lu-N3 2.371(3), P1-N1 1.609(3), P1-N2 1.610(4), P2-N3 1.608(4); N1-Lu-N1' 156.7(2), N2-Lu-N1 62.95(12), N3-Lu-N3' 63.1(2), N2-P1-N1' 100.9(2), N3-P2-N3' 101.0(2).

3.7 Enantiopure group 14 compounds

3.7.1 Introduction

In low-valent main group chemistry, amido (NR_2^-) ligands play an important role in order to stabilize a variety of elusive species.[47] Since the discovery of the first isolable N-heterocyclic carbene (NHC) by Arduengo *et al.*[159], NHCs have been widely employed as ancillary ligands in metal-mediated catalysis,[160] material science,[161] and, most recently, biologically relevant complexes.[162]

The field of NHC metal complexes is not only limited to achiral catalysis, but also several chiral NHCs have been reported and utilized to impart enantioselectivity in organic transformations.[163] In contrast to the well-developed chemistry of chiral NHCs[164] and their successful applications in organocatalysis,[165] the corresponding chemistry with tetrylenes, which are heavier divalent analogues of group 14 elements, is limited to a few ligand backbones.[166]

Herein, as a preliminary study, chiral tetrylenes are presented. To impart chirality in the compounds and as a comparative analysis, chiral iminophosphonamide and amidinate ligands, as depicted in Chart 3.7.1, have been utilized.

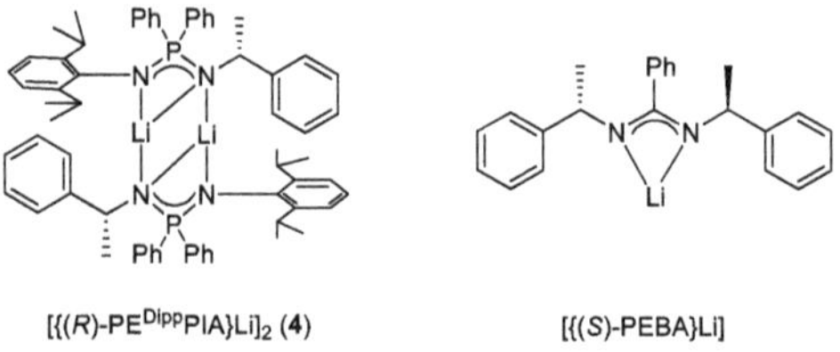

[{(*R*)-PE^{Dipp}PIA}Li]$_2$ (**4**)   [{(*S*)-PEBA}Li]

Chart 3.7.1 Selected iminophosphonamide and amidinate ligands to synthesize low valent group 14 compounds.

3.7.2 Synthesis of chiral tetrylenes

In order to introduce chirality into the benzamidinato-supported silylene [{PhC(^{t}BuN)$_2$SiCl], [{(*R*)-PE^{Dipp}PIA}Li]$_2$ (**4**) was reacted with [{PhC(^{t}BuN)$_2$}SiCl] in a 1:2 molar ratio. Interestingly, the expected salt metathesis product was not formed but a formally Si(+IV) compound, [{PhC(N^{t}Bu)$_2$}Si{N(*R*)(CH)MePh(PPh$_2$)}{=NDipp}] (**36**), was isolated instead (Scheme 3.7.1).

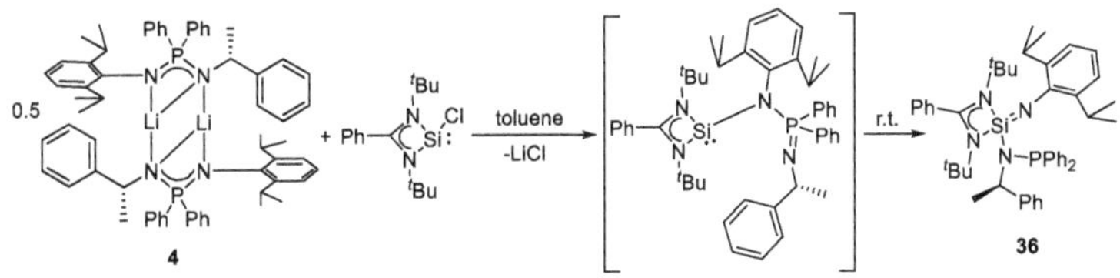

Scheme 3.7.1 Synthesis of [{PhC(N^{t}Bu)$_2$}Si{N(*R*)(CH)MePh(PPh$_2$)}{=NDipp}] (**36**).

The formation of the unexpected compound **36** can be rationalized by *in situ* activation of one P-N bond in the {(*R*)-PE^{Dipp}PIA}$^-$ moiety through the suggested silylene intermediate (Scheme 3.7.1). This intermediate silylene species undergoes an oxidative addition leading to the formation of the chiral silane **36**. The reactivity of silylenes towards iminophosphonamide compounds has not been reported beforehand. However, a similar reactivity of silylenes (*i.e.* [{PhC(^{t}BuN)$_2$}SiCl]) with phosphonamides has been reported by the group of So.[167] More recently, our group has reported the analogous activation of an As-N bond by the same silylene.[168] In the ^{1}H NMR spectrum, a characteristic septet resonance for the C*H*(CH$_3$)$_2$ groups is observed at δ= 4.18 ppm ($^3J_{HH}$ = 6.8 Hz). The corresponding methyl protons CH(C*H*$_3$)$_2$ appear as two distinct doublets at δ= 1.43 ppm ($^3J_{HH}$ = 6.8 Hz) and δ= 1.37 ppm ($^3J_{HH}$ = 6.8 Hz) due to restricted rotation. The resonances for the Ph(C*H*)CH$_3$ and Ph(CH)C*H*$_3$ protons can be detected as a quartet and a doublet at δ= 5.63 ppm ($^3J_{HH}$ = 7.28 Hz) and δ= 2.14 ppm ($^3J_{HH}$ = 7.28 Hz), respectively. A singlet for C(C*H*$_3$)$_3$ is observed at δ = 1.18 ppm. The ^{31}P{^{1}H} NMR spectrum exhibits a singlet at δ = 50.6 ppm, significantly shifted downfield as compared to the resonance at δ= 2.9 ppm in case of compound **4**. Furthermore, in the ^{29}Si{^{1}H} NMR spectrum, a doublet at δ = -83.5 ppm ($^2J_{P\text{-}Si}$ = 48.1 Hz) is observed, which is significantly shifted upfield when compared to δ= 14.6 ppm for [{PhC(^{t}BuN)$_2$}SiCl].[15b]

The solid state structure of compound **36** was determined by X-ray diffraction studies. Compound **36** crystallizes in the orthorhombic chiral space group $P2_12_12_1$ (Figure 3.7.1). The crystal structure of **36** shows that the Si-N1 bond length of 1.865(5) Å is slightly longer than Si-N2 1.839(6) Å, which may be attributed a minimization of steric repulsion between the Dipp-group and the tertiary-butyl-group. The Si-N3 bond length (1.569(5) Å) is shorter and is within the reported range (1.545(2)-1.589(3) Å) of Si=N double bonds.[168-169] The silicon atom in **36** adopts a distorted tetrahedral geometry with a N1-Si-N2 bite angle of 71.1(2)°. The P-N bond length of 1.718(5) Å lies in the range of reported P-N single bond lengths.[170]

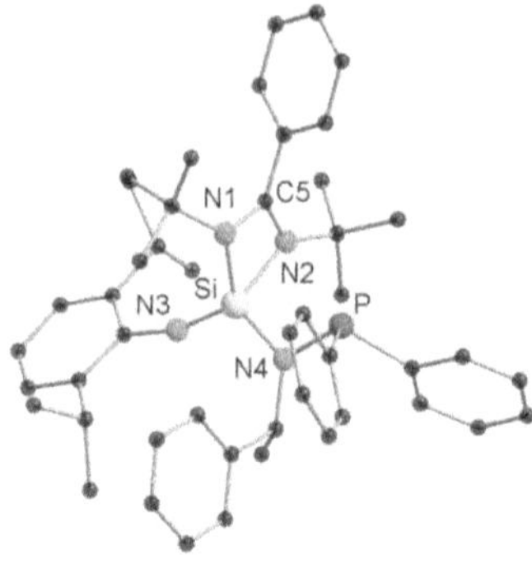

Figure 3.7.1 Molecular structure of **36** in the solid state. All hydrogen atoms are omitted for clarity. Selected bond lengths (Å) and bond angles [°]: Si-N1 1.865(5), Si-N2 1.839(6), Si-N3 1.569(5), Si-N4 1.745(5), N1-C5 1.324(8), N2-C5 1.342(8), P-N4 1.718(5); Si-N1-C5 90.2(4), Si-N2-C5 90.8(4), N1-C5-N2 107.8(5), N3-Si-N4 118.0(3), P-N4-Si 114.1(3), N1-Si-N2 71.1(2), N1-Si-N4 105.6(2), N1-Si-N3 124.5(3), N2-Si-N4 111.5(2).

To further investigate this unexpected reactivity, synthesis of the heavier germylene compound was attempted by reaction of $GeCl_2$·dioxane with complex **4** in a 2:1 molar ratio, as depicted in (Scheme 3.7.2). The reaction proceeded smoothly with the facile elimination of LiCl and led to the formation of the expected compound **37**. Favourable formation of **37** is in agreement with the lower reactivity of germylenes compared to silylenes.[171] Owing to the inter pair effect, low oxidation states are more stable moving down the group. Compound **37** is a colourless crystalline solid, highly soluble in polar organic solvents such as thf, toluene and Et_2O, whereas, moderately soluble in *n*-pentane. Compound **37** was further characterized by various analytical techniques such as mass spectrometry, elemental analysis, IR, and single crystal X-ray studies. Compound **37** slowly decomposes in solution and therefore clean NMR spectra could not be recorded. However, in the EI mass spectrum of **37**, the molecular ion peak was observed at m/z = 588.149, which corresponds to $[M]^+$, and another signal at m/z 553.170 for $[M-Cl]^+$.

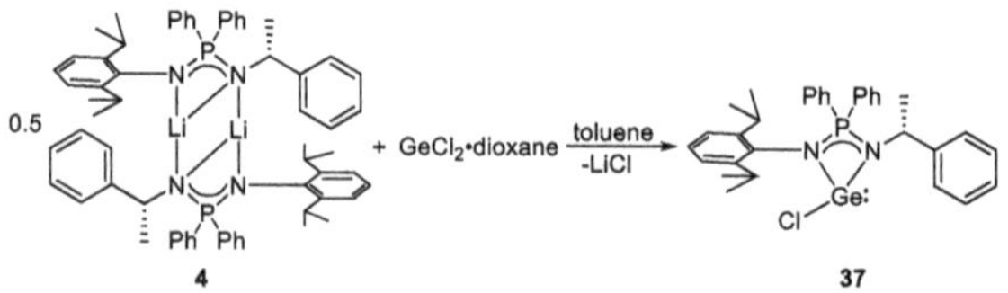

0.5 **4** + $GeCl_2$•dioxane $\xrightarrow[\text{-LiCl}]{\text{toluene}}$ **37**

Scheme 3.7.2 Synthesis of [{(*R*)-PE^{Dipp}PIA}GeCl] (**37**).

The molecular structure of **37** was determined by single crystal X-ray diffraction analysis. It crystallises in the orthorhombic chiral space group $P2_12_12_1$.

The molecular structure of compound **37** shows that the Ge(II) centre adopts a trigonal pyramidal geometry, coordinated by two nitrogen centres (N1 and N2) of {(*R*)-PE^{Dipp}PIA}$^-$ and one chlorine atom (Figure 3.7.2). The Ge–Cl bond length (2.3169(11) Å) is longer as compared to that in the amidinate compounds [PhC(N^{t}Bu)$_2$]GeCl (2.257(2) Å)[15c] and [^{t}BuC(N-2,6-iPr$_2$C$_6$H$_3$)$_2$]GeCl (2.174(2) Å).[172] It is however similar to that observed with the achiral iminophosphonamide compound [(2,6-iPr$_2$C$_6$H$_3$N)P(Ph$_2$)(N^{t}Bu)]GeCl (2.338(1) Å).[49a]

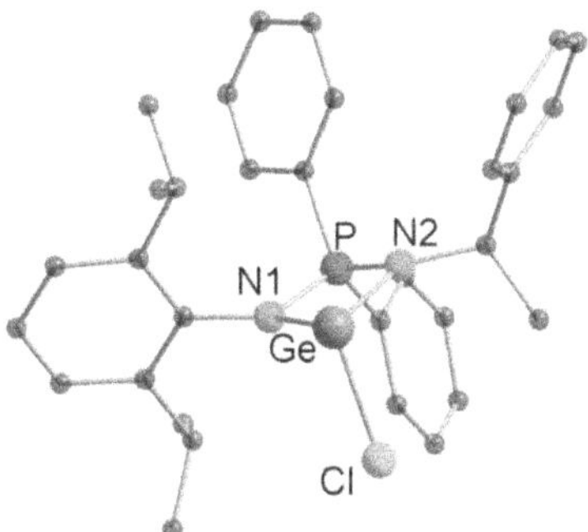

Figure 3.7.2 Molecular structure of **37** in the solid state. All hydrogen atoms are omitted for clarity. Selected bond lengths (Å) and bond angles [°]: Ge-Cl 2.3169(11), Ge-N1 2.010(2), Ge-N2 2.012(2), P-N1 1.617(2), P-N2 1.607(3); N1-Ge-N2 72.77(10), N1-Ge-Cl 96.68(8), N2-Ge-Cl 100.36(8), N1-Ge-P 36.52(7), N2-Ge-P 36.26(7), Cl-Ge-P 99.81(4).

Furthermore, the Ge-N bond lengths in **37** (Ge-N1 2.010(2), Ge-N2 2.012(2) Å) are slightly shorter than in [PhC(N*t*Bu)$_2$]GeCl (2.060(2) Å), whereas longer than those observed in [(2,6-iPr$_2$C$_6$H$_3$N)P(Ph$_2$)(N^{t}Bu)]GeCl (1.980(3) and 1.992(3) Å).

The N1-Ge-N2 bite angle is wider as compared to the corresponding angle in [PhC(N^{t}Bu)$_2$]GeCl (63.22(11)°). However, it is slightly narrower than that in [(2,6-iPr$_2$C$_6$H$_3$N)P(Ph$_2$)(N^{t}Bu)]GeCl (73.08(2)°). The central germanium centre lies slightly above to the NPN plane with the dihedral angle of 1.97° between the N1-P-N2 and N1-Ge-N2 planes.

Due to presence of a P(V) reducible centre in the ligand backbone of {(R)-PE^{Dipp}PIA}$^-$, the attempted synthesis of a chiral silylene led to the formation the chiral silane **36**. Therefore, in order to introduce chirality, the (*S*)-phenyl ethyl benzamidinate {(*S*)-PEBA}$^-$ ligand was chosen, since it does not possess any redox-active centre. The reaction of [{PhC(^{t}BuN)$_2$}SiCl] with [{(*S*)-

PEBA}Li] in an equimolar ratio led to the formation of [{PhC(^{t}BuN)$_2$}{(*S*)-PEBA}Si] (**38**) (Scheme 3.7.3).

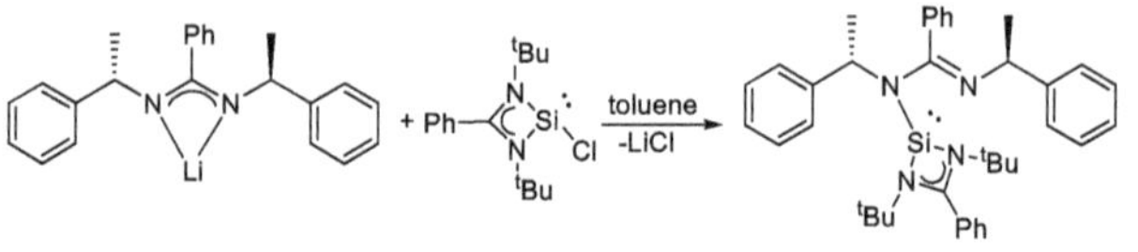

[{(*S*)-PEBA}Li] + Ph–C(^{t}BuN)$_2$SiCl —(toluene, -LiCl)→ **38**

Scheme 3.7.3 Synthesis of [{PhC(^{t}BuN)$_2$}{(*S*)-PEBA}Si] (**38**).

The identity of compound **38** was confirmed by characterization *via* NMR, elemental analysis, IR and single crystal X-ray diffraction studies. The ^{1}H NMR spectrum of **38** exhibits a complex set of signals, suggesting the nonsymmetric nature of the compound. The Ph(C*H*)CH$_3$ protons appear as two different quartets at δ= 4.82 ppm ($^3J_{HH}$ = 6.83 Hz) and δ= 4.43 ppm ($^3J_{HH}$ = 6.42 Hz). Similarly, the methyl Ph(CH)C*H*$_3$ protons appear as two different doublets at δ= 2.41 ppm ($^3J_{HH}$ = 6.87 Hz) and δ= 1.34 ppm ($^3J_{HH}$ = 6.38 Hz). Due to the nonsymmetric nature of **38**, the C(C*H*$_3$)$_3$ groups appear as two distinct singlets at δ= 1.09 ppm and δ= 0.90 ppm. Furthermore, the ^{29}Si{^{1}H} NMR exhibits one single resonance at δ= -19.1 ppm, which is significantly shifted upfield as compared to the signal at δ = 14.6 ppm for [{PhC(^{t}BuN)$_2$}SiCl],[15b] whereas it is significantly shifted downfield as compared to the resonance at δ= -83.5 ppm for compound **36**.

Compound **38** crystallizes in the chiral monoclinic space group $P2_1$ (Figure 3.7.3). The central silicon atom adopts a trigonal pyramidal geometry. The coordination sphere around the silicon centre is made up of one nitrogen atom from the {(*S*)-PEBA}$^-$ ligand and two nitrogen atoms from the [{PhC(^{t}BuN)$_2$}$^-$ ligand. The Si-N bond lengths in **38** (Si-N1 1.870(2) Å, Si-N2 1.880(2) Å, Si-N3 1.806(2) Å) are in the range of Si-N single bonds.[173] The N1-C9 and N2-C9 bond lengths are 1.338(3) Å, 1.344(3) Å, respectively, in between the range for N-C single and double bonds. However the N3-C24 (1.384(3) Å) and N4-C24 (1.281(3) Å) show a characteristic single and double bond character, respectively.[174]

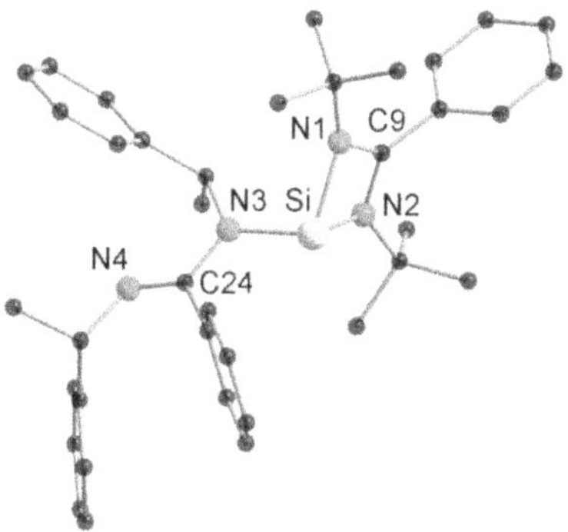

Figure 3.7.3 Molecular structure of **38** in the solid state. All hydrogen atoms are omitted for clarity. Selected bond lengths (Å) and bond angles [°]: Si-N1 1.870(2), Si-N2 1.880(2), Si-N3 1.806(2), N1-C9 1.338(3), N2-C9 1.344(3), N3-C24 1.384(3), N4-C24 1.281(3); N1-Si-N2 69.16(9), N1-Si-N3 102.16(10), N2-Si-N3 102.68(10), N1-C9-N2 105.1(2), C24-N3-Si 118.8(2), N3-C24-N4 120.8(2).

Owing to chelation of the Si atom, the N1-C9-N2 angle (105.1(2)°) is narrower than the N3-C24-N4 (120.8(2)°) bond angle. Encouraged by the successful isolation of compound **38**, the respective chiral germylene was also synthesized. As expected, the reaction of [{(*S*)-PEBA}Li] with $GeCl_2$·dioxane in an equimolar ratio resulted in compound **39** (Scheme 3.7.4). Due to a nonsymmetric coordination environment, two distinct quartets at δ = 4.35 ppm ($^3J_{HH}$ = 6.76 Hz), and δ = 4.22 ppm ($^3J_{HH}$ = 6.77 Hz) for Ph(C*H*)CH_3 are observed in the ^{1}H NMR spectrum. Furthermore, the Ph(CH)C*H*$_3$ protons appear as doublets at δ = 1.45 ppm ($^3J_{HH}$ = 6.77 Hz) and δ = 1.23 ppm ($^3J_{HH}$ = 6.79 Hz). The nonsymmetric character was further confirmed by the appearance of two different sets of signals for the Ph(*C*H)CH_3 at δ = 56.6 and 56.5 ppm and Ph(CH)CH_3 at δ = 25.8 and 24.3 ppm in the ^{13}C{^{1}H} NMR spectrum. Single crystals of compound **39** suitable for X-ray analysis were obtained from hot *n*-heptane.

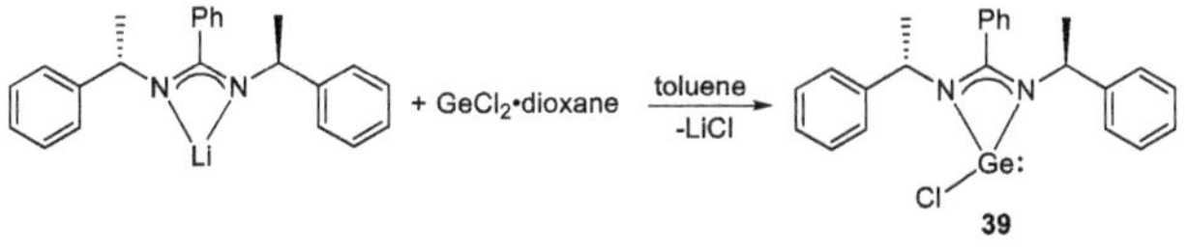

Ph N N Li + $GeCl_2$•dioxane —toluene / -LiCl→ Ph N N Ge: Cl **39**

Scheme 3.7.4 Synthesis of compound [{(*S*)-PEBA}GeCl] (**39**).

Compound **39** crystallizes in orthorhombic chiral $P2_12_12_1$ space group with one molecule in the asymmetric unit (Figure 3.7.4). The Ge-Cl bond length in **39** is 2.2690(10) Å, which is shorter as compared to 2.3169(11) Å in compound **37**, whereas it is in the reported range for similar amidinate compounds such as [{PhC(N^{t}Bu)$_2$}GeCl] (2.257(2) Å)[15c] and [^{t}BuC(N-2,6-

$iPr_2C_6H_3)_2\}GeCl]$ (2.174(2) Å).[172] Furthermore, the N1-C1 and N2-C1 bond lengths are of equal lengths (1.331(4) Å), suggesting complete delocalisation of the anionic charge. The N1-Ge-N2 bite angle in compound **39** (65.59(11)°) is narrower than that in **37** (72.77(10)°). Furthermore, the germanium centre shows a slight deviation from the NCN plane with a dihedral angle of 10.9(2)°. Almost equal bond angles for N1-Ge-C1 (32.98(10)°) and N2-Ge-C1 (32.93(11)°) suggest that the germanium centre lies exactly in the bisection of the N1-Ge-N2 angle.

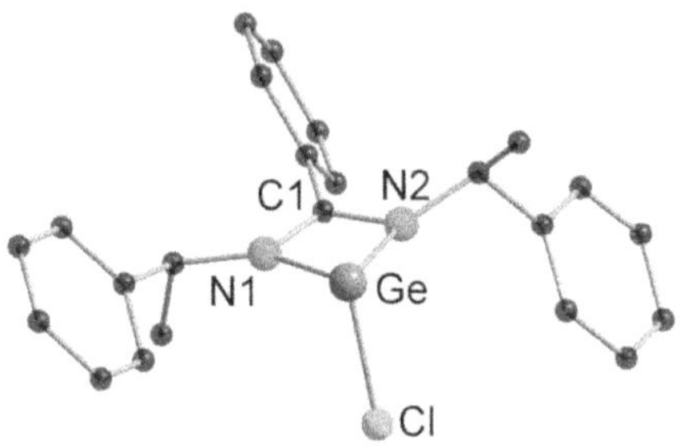

Figure 3.7.4 Molecular structure of **39** in the solid state. All hydrogen atoms are omitted for clarity. Selected bond lengths (Å) and bond angles [°]: Ge-Cl 2.2690(10), Ge-N1 2.021(3), Ge-N2 1.987(3), N1-C1 1.331(4), N2-C1 1.331(4); N1-Ge-Cl 96.18(8), N2-Ge-Cl 97.63(10), N1-Ge-N2 65.59(11), N1-Ge-C1 32.98(10), N2-Ge-C1 32.93(11), N1-C1-N2 109.3(3).

4 Experimental Section

4.1 General consideration

All manipulations of air and moisture sensitive materials were performed under the rigorous exclusion of oxygen and moisture in flame-dried Schlenk-type glassware either on a dual manifold Schlenk line, interfaced to a high vacuum (10^{-3} torr) line, or in an argon-filled MBraun glove box. Hydrocarbon solvents (toluene, diethylether, *n*-pentane, *n*-heptane) were dried by using an MBraun solvent purification system (SPS-800), degassed and stored under vacuo over lithium aluminium hydride ($LiAlH_4$). *n*-hexane was pre-dried over $CaCl_2$ before decantation and distillation from potassium and storage over 4 Å molecular sieves. Tetrahydrofuran was distilled under nitrogen from potassium benzophenoneketyl before storage over lithium aluminium hydride ($LiAlH_4$). NMR spectra were recorded on a Bruker NMR spectrum were recorded on a BrukerAvance II 300 MHz or Avance 400 MHz. Chemical shifts are expressed in parts per million (ppm) and referenced on characteristic solvent resonances as internal standards [C_6D_6: 7.16 ppm (^1H) and 128.06 ppm (^{13}C); $CDCl_3$: 7.26 ppm (^1H) and 77.16 ppm (^{13}C)]; and are reported relative to tetramethyl silane. 85% phosphoric acid was used as external reference for ^{31}P- and $^{31}P\{^1H\}$-NMR. ^1H-NMR are reported as follows: chemical shift (*d* in ppm), multiplicity (br for broad singlet, s for singlet, d for doublet, q for quartet, m for multiplet), coupling constant(s) (Hz), number of protons (concluded from the integrals), specific assignment. $^{13}C\{^1H\}$ NMR spectra are reported in terms of chemical shift and specific assignment. NMR assignments were made using a combination of 1D and 2D techniques [^1H-^1H COSY, ^1H-^{13}C HMQC, and ^1H-^{13}C HMBC]. Elemental analyses were carried out with a Vario Micro Cube (Elementar Analysensysteme GmbH). IR spectra were obtained on a Bruker Tensor 37 FTIR spectrometer equipped with a room temperature DLaTGS detector, a diamond ATR (attenuated total reflection) unit, and a nitrogen flushed chamber. In terms of their intensity, the signals were classified into the categories vs = very strong, s = strong, m = medium, w = weak and vw = very weak.

4.2 Synthesis of Starting Materials

The amines (*R*)-α-methylbenzylamine, (*R*)-α-methylnaphthylamine and Dipp amine (Dipp = 2,6-$^{i}Pr_2C_6H_3$) were purchased from commercial sources and dried by CaH_2 and distilled before use.[175]

For synthesis of (*R*)-α-methylnaphthyl azide the literature procedure was exactly followed.[107] Dipp-azide (Dipp = 2,6-$^{i}Pr_2C_6H_3$),[104] Ph_2PN(*R*-*CHMePh),[105] Ph_2PN(*R*-*CHMeNaph),[106] *P,P*-diphenyl-*N,N'*-bis((*R*)-1-phenylethyl)phosphinimidic amine, (*R*)-HPEPIA,[103b] Rb[N($SiMe_3)_2$],[176] Cs[N($SiMe_3)_2$],[177] [MgN($SiMe_3)_2$],[178] [CaN($SiMe_3)_2$(thf$)_2$],[126] and *P,P*-diphenyl-*N,N'*-bis(2,6-diisopropyl phenyl)phosphinimidic amine, HDippPIA[26], [{(*S*)-PEBA}]-Li,[48b] and [(PhC($N^tBu)_2$)SiCl],[48a] were also prepared according to already described procedures. The other chemicals were purchased from commercial sources and used without further purification if not specified.

Note: NMR studies of complexes **33**, **34** and **37** were not possible due to their paramagnetic character for first two complexes and rapid decomposition in solution for **37**.

4.3 Synthesis and analytical data for chiral iminophosphonamines and its alkali metal complexes

4.3.1 Synthesis of (*R*)-$HPE^{Dipp}PIA$

Ph_2PN(*R*-*CHMePh) (3.2 g ,10.61 mmol, 1.0 eq) was dissolved in 30 mL of thf and cooled to -20 °C. A solution of Dipp-azide (2.4 g, 11.67 mmol, 1.1 eq) in 20 mL of thf was added slowly. After complete addition, the reaction mixture was warmed to ambient temperature whereupon liberation of N_2 gas was observed. The reaction mixture was stirred for 16 hours at ambient temperature. After evaporation of the solvent under reduced pressure, a viscous solid was obtained, which was washed with *n*-pentane (4*20 mL) and dried in vacuum to obtain the desired product as white powder. Single crystals of the title compound suitable for X-ray analysis were obtained by recrystallization from hot *n*-heptane.

Yield (based on crystals): 3.57 g (70 %) **Elemental analysis** calcd. (%) for [$C_{32}H_{37}N_2P$] (480.64): C 79.97, H 7.76, N 5.83; found: C 79.42, H 7.63, N 5.80.

^1H NMR (C_6D_6, 400 MHz): δ [ppm] = 7.78-7.66 (m, 4 H, *o*-Ar_{phos}-*H*), 7.25-7.22 (m, 2 H, Ar-*H*), 7.10-6.92 (m, 12 H *o, m ,p*-Ar-*H*), 4.64 (ddq, $^3J_{HH}$ = 6.8 Hz, $^3J_{HH}$ = 10 Hz, $^3J_{PH}$ = 10 Hz 1 H, HNC*H*), 3.67 (sept, $^3J_{HH}$ = 6.9 Hz, 2 H, *H*C($CH_3)_2$), 2.81 (dd, $^3J_{HH}$ = 9.5 Hz, $^2J_{PH}$ = 9.5 Hz 1 H, N*H*), 1.26 (d,

$^3J_{HH}$ = 6.8 Hz, 3 H, HNCHC*H*$_3$) , 1.183 (d, $^3J_{HH}$ = 6.9 Hz, 6 H, HC(C*H*$_3$)$_2$), 1.180 (d, $^3J_{HH}$ = 6.9 Hz, 6 H, HC(C*H*$_3$)$_2$).

^{13}C{^{1}H} NMR (C_6D_6, 100 MHz): δ [ppm] = 146.2 (d, $^3J_{PC}$ = 4.8 Hz, PNHCHC$_q$), 144.6 (Ar-C_{DiP}), 142.0 (d, $^2J_{PC}$ = 7.2 Hz P=NC$_q$), 134.8 (d, $^1J_{PC}$ = 121 Hz, Ar$_{phos}$C$_q$), 134.2 (d, $^1J_{PC}$ = 127 Hz, Ar$_{phos}$C$_q$), 132.3 (d, $^2J_{PC}$ = 9.8 Hz *o*-Ar$_{phos}$-*C*H), 132.2 (d, $^2J_{PC}$ = 9.8 Hz, *o*-Ar$_{phos}$-*C*H), 131.1 (d, $^3J_{PC}$ = 2.8 Hz, Ar-*C*H), 130.9 (d, $^3J_{PC}$ = 2.9 Hz, Ar-*C*H), 128.7 (Ar-*C*), 128.5 (Ar-*C*), 128.2 (Ar-*C*), 127.0 (Ar-*C*), 126.2 (Ar-*C*), 123.3 (d, $^4J_{PC}$ = 1.3 Hz, *o*-Ar-*CH*), 119.9 (d, J_{PC} = 2.5 Hz, Ar-*C*), 50.8 (P-N*C*HCH$_3$), 29.1 (*C*H(CH$_3$)$_2$), 26.1 (d, $^3J_{PC}$ = 4.0 Hz, Ph(CH)*C*H$_3$), 24.13 (CH(*C*H$_3$)$_2$), 24.06 (CH(*C*H$_3$)$_2$).

^{31}P{^{1}H} NMR (C_6D_6, 162 MHz): δ [ppm] = -10.9.

IR (ATR): $\tilde{\nu}$ [cm^{-1}] = 3347 (w), 3054 (vw), 2971 (w), 2961 (vw), 2864 (w), 1585 (w), 1596 (w), 1450 (m), 1434 (vs), 1401 (vw), 1354 (m), 1295 (w), 1257 (w), 1202 (w), 1185 (w), 1117 (m), 1101 (m), 1087 (m), 1075 (m), 1064 (m), 1028 (w), 1017 (w), 950 (m), 833 (w), 794 (vw), 771 (m), 749 (vs), 695(vs), 625 (s), 649 (vw), 600 (m), 534 (s), 511 (vs), 500 (vs), 447 (w).

4.3.2 Synthesis of (*R*)-HNEPIA

Ph_2PN(*R*-*CHMeNaph) (7.20 g, 20.26 mmol, 1.0 eq.) were dissolved in thf (40 ml) and cooled to -20 °C. A solution of 2.37 g (*R*)-α-methylnaphthyl azide (22.28 mmol, 1.10 eq.) in thf (20 ml) was added slowly. After complete addition the reaction mixture was warmed to ambient temperature whereupon liberation of N_2 was observed. The reaction mixture was stirred for 12 h at ambient temperature. After evaporation of the solvent under reduced pressure, a viscous solid was obtained, which was washed with *n*-pentane (4*20 mL) and dried in vacuum to obtain the desired product as white powder. Single crystals of the title compound suitable for X-ray Analysis were obtained by recrystallization from *n*-heptane.

Yield (based on crystals): 6.7 g (63 %) **Elemental analysis** calcd. (%) for [$C_{36}H_{33}N_2P$] (524.65): C 82.42, H 6.34, N 5.34; found: C 82.10, H 6.23, N 5.40.

^{1}H NMR (C_6D_6, 400 MHz): δ [ppm] = 8.44 (br, 2 H. Ar-*H*), 8.04 (br, 2 H, Ar-*H*), 7.77-7.69 (br, 3 H, Ar-*H*), 7.52-7.19 (m, 10 H, Ar-*H*), 7.09-6.95 (m, 7 H, Ar-*H*), 5.52-5.45 (m, 1 H, P=NC*H*), 5.17 (br, 1 H, HNC*H*), 2.88 (br, 1 H, N*H*), 1.81 (br, 3 H, P=NCHC*H*$_3$), 1.21 (br, 3 H, HNCHC*H*$_3$).

^{13}C{^{1}H} NMR (C_6D_6, 100 MHz): δ [ppm] = 148.0 (d, J_{PC} = 11.3 Hz, Ar-*C*), 142.6 (Ar-*C*), 135.6 (d, J_{PC} =23.5 Hz, Ar-*C*), 134.5 (Ar-*C*), 134.2 (Ar-*C*H), 132.9 (d, J_{PC} =8.2 Hz, Ar-*C*H), 132.1 (d, J_{PC} =8.6

Hz, Ar-*C*H), 131.4 (Ar-*C*H), 130.8 (Ar-*C*H), 129.0 (d, J_{PC} =20.5 Hz, Ar-*C*H), 127.9 (Ar-*C*H), 127.6 (Ar-*C*H), 126.5 (Ar-*C*H), 126.1 (d, J_{PC} = 8.2 Hz, Ar-*C*H), 125.5 (d, J_{PC} =13.1 Hz, Ar-*C*H), 125.0 (d, J_{PC} =16.1 Hz, Ar-*C*H), 124.7 (Ar-*C*H), 124.3 (Ar-*C*H), 123.4 (Ar-*C*H), 122.4 (Ar-*C*H), 51.2 (P=N*C*H), 46.3 (HN*C*H), 29.9 (d, $^3J_{PC}$ = 11.8 Hz, P=NCH*C*H$_3$), 25.7 (HNCH*C*H$_3$).

^{31}P{^{1}H} NMR (C_6D_6, 162 MHz): δ [ppm] = 3.6.

IR (ATR): $\tilde{\nu}$ [cm^{-1}] = 3372 (m), 3074 (m),3051 (m) 2965 (m), 2921 (w), 2858 (vw), 2826 (vw), 2166 (vw), 2104 (vw), 1594 (m), 1508 (m), 1480 (vw), 1435 (m), 1392 (s), 1374 (m), 1329 (m), 1287 (s) 1261 (m), 1233 (vs), 1165 (s), 1111 (s), 1102 (m), 1080 (w), 1053 (w), 1024 (w),997 (w), 934 (m), 868 (w), 849 (w), 829 (s), 801 (m), 774 (s), 750 (s), 717(m), 694 (m), 639 (m), 612 (w), 545 (w), 524 (m), 483 (w), 434 (w).

4.3.3 Synthesis of [{(*R*)-PEPIA}$_2$K$_2$] (1)

(*R*)-HPEPIA (4.47 g, 10.5 mmol, 1.0 eq.) and potassium hydride (60 % in mineral oil) (506 mg, 12.7 mmol, 1.20 eq.) were dissolved in thf (40 mL) at room temperature. The resulting reaction mixture was stirred overnight at room temperature. The colour of the reaction mixture changed from yellow to dark red. After extracting of the reaction mixture with toluene and washing with *n*-pentane, the light-yellow solid was recrystallised from hot toluene to obtain colourless crystals, suitable for X-ray structure analysis. The solvent was decanted, and the product was washed one time with *n*-pentane (5 mL).

Yield (based on crystals): 1.60 g (33 %). **Elemental analysis** calcd. (%) for [$C_{56}H_{56}K_2N_4P_2$] (925.21): C 72.70, H 6.10, N 6.06; found: C 72.84, H 5.80, N 6.06.

^{1}H NMR (thf-d$_8$, 300 MHz): δ [ppm] = 7.52-7.46 (m, 8 H, *o*-Ar$_{phos}$-*C*H), 7.31-7.29 (m, 8 H, *o*-*C*H), 7.14-7.05 (m, 20 H, *m*-*CH* & *m,p*-Ar$_{phos}$-CH), 6.97-6.91 (m, 4 H, *p*-*C*H), 4.05 (dq, $^3J_{HH}$ = 6.3 Hz, $^3J_{PH}$ = 23.4 Hz, 4 H, *C*H), 0.92 (d, $^3J_{HH}$ = 6.4 Hz, 12 H, *C*H$_3$).

^{13}C{^{1}H} NMR (thf-d$_8$, 75 MHz): δ [ppm] = 155.4 (d, $^3J_{PC}$ = 14.1 Hz, Ar-*C*$_q$), 142.4 (d, $^1J_{PC}$ = 76.9 Hz, Ar$_{phos}$-*C*$_q$), 133.0 (d, $^2J_{PC}$ = 7.7 Hz, *o*-Ar$_{phos}$-*C*H), 128.2 (d, $^4J_{PC}$ = 2.5 Hz, *o*-Ar-*C*H), 128.0 (*m*-*C*H), 127.4 (*o*-*C*H), 127.3 (d, $^3J_{PC}$ = 9.4 Hz, *m*-Ar$_{phos}$-*C*H), 125.0 (*p*-*C*H), 55.3 (d, $^2J_{PC}$ = 2.0 Hz, *C*H), 30.4 (d, $^3J_{PC}$ = 8.7 Hz, *C*H$_3$).

^{31}P{^{1}H} NMR (thf-d$_8$, 121 MHz): δ [ppm] = 16.3.

IR (ATR): $\tilde{\nu}$ [cm^{-1}] = 3050 (vw), 3022 (vw), 2962 (vw), 2913 (vw), 2852 (vw), 1597 (vw), 1489 (w), 1479 (w), 1431 (w), 1358 (vw), 1345 (w), 1314 (w), 1274 (w), 1171 (m), 1147 (s), 1138 (m), 1101 (w), 1063 (w), 1027 (w), 1003 (vw), 977 (w), 906 (vw), 808 (m), 781 (vw), 756 (s), 742 (vs), 709 (w), 698 (s), 620 (s), 608 (s), 543 (s), 522 (s), 498 (s), 479 (vw).

Raman (solid state): $\tilde{\nu}$ [cm^{-1}] = 3050 (m), 2973 (vw), 2914 (vw), 2855 (vw), 1587 (w), 1196 (m), 1180 (vw), 1154 (vw), 1100 (w), 1028 (vw), 1003 (w), 997 (w), 982 (vs), 816 (vw), 760 (vw), 676 (vw), 621 (vw), 285 (vw), 230 (vw), 92 (w).

4.3.4 Synthesis of [{(*R*)-PEPIA}$_2$Rb$_2$] (2)

Path A: (*R*)-HPEPIA (346 mg, 0.81 mmol, 1.0 eq.) and 200 mg of $RbN(SiMe_3)_2$ (0.81 mmol, 1.0 eq.) were dissolved in thf (10 mL) and stirred over three days at room temperature. The reaction mixture was filtered and the solvent was evaporated under reduced pressure. The resulting residue was washed with *n*-pentane (3*5 mL) which resulted in a white microcrystalline solid. Single crystals suitable for X-ray analysis were obtained by recrystallization from *n*-hexane.

Yield (based on crystals): 256mg (62 %). **Elemental analysis** calcd. (%) for [$C_{56}H_{56}Rb_2N_4P_2$] (1017.95): C 66.07, H 5.54, N 5.50; found: C 66.68, H 5.55, N 5.67.

Path B: (*R*)-HPEPIA (298 mg, 0.70 mmol, 1.0 eq.) and 72 mg of elemental rubidium (0.84 mmol, 1.20 eq.) were placed in a Schlenk flask. Toluene (20 mL) was added and the mixture was stirred for 3 h at 90 °C. The reaction mixture was cooled to room temperature. After filtration and removal of all volatiles, the off-white residue was recrystallized from *n*-hexane. The product was obtained in form of colourless prismatic shaped crystals. **Yield** (based on crystals): 166 mg (39 %).

^{1}H NMR (thf-d8, 300 MHz): δ[ppm] = 7.54-7.49 (m, 8 H, *o*-Ar$_{phos}$-C*H*), 7.28-7.26 (m, 8 H, *o*-C*H*), 7.14-7.05 (m, 20 H, *m-CH* & *m, p*-Ar$_{phos}$-CH), 6.96-6.91 (m, 4 H, *p*-C*H*), 4.06 (dq, $^3J_{HH}$ = 6.4 Hz, $^3J_{PH}$ = 23.4 Hz, 4 H, C*H*), 0.96 (d, $^3J_{HH}$ = 6.4 Hz, 12 H, C*H*$_3$).

^{13}C{^{1}H} NMR (thf-d$_8$, 75 MHz): δ [ppm] = 155.4 (d, $^3J_{PC}$ = 14.4 Hz, Ar-C_q), 142.7 (d, $^1J_{PC}$ = 78.0 Hz, Ar$_{phos}$-C_q), 133.0 (d, $^2J_{PC}$ = 7.6 Hz, *o*-Ar$_{phos}$-*C*H), 128.2 (s, *p*-Ar$_{phos}$-CH), 128.0 (*m*-CH), 127.4 (s, *o*-Ar-*C*H), 127.3 (d, $^3J_{PC}$ = 9.3 Hz, *m*-Ar$_{phos}$-CH), 125.0 (*p*-*C*H), 55.6 (s, *C*H), 30.2 (d, $^3J_{PC}$ = 8.7 Hz, *C*H$_3$).

$^{31}P\{^1H\}$ NMR (thf-d_8, 121 MHz): δ [ppm] = 14.7.

IR (ATR): $\tilde{\nu}$ [cm^{-1}] = 3528 (vw), 3054 (vw), 3022 (vw), 2960 (vw), 2911 (vw), 2851 (vw), 1596 (vw), 1489 (w), 1449 (w), 1433 (w), 1358 (vw), 1345 (w), 1313 (w), 1274 (w), 1168 (m), 1147 (s), 1137 (m), 1100 (w), 1063 (w), 1027 (w), 1002 (vw), 977 (w), 945 (w), 907 (vw), 807 (m), 781 (vw), 756 (s), 745 (vs), 696 (s), 620 (s), 607 (s), 543 (s), 520 (s), 497 (s), 478 (vw).

Raman (solid state): $\tilde{\nu}$ [cm^{-1}] = 3048 (m), 2973 (vw), 2912 (vw), 2853 (vw), 1598 (w), 1587 (w), 1196 (m), 1180 (vw), 1155 (vw), 1099 (w), 1028 (vw), 1003 (w), 998 (w), 979 (vs), 813 (vw), 777 (vw), 761 (vw), 676 (vw), 621 (vw), 285 (vw), 230 (vw), 92 (w).

4.3.5 Synthesis of [{(*R*)-PEPIA}$_2$Cs$_2$]/[{(*R*)-PEPIA}Cs]$_n$ (3)

Path A: (*R*)-HPEPIA (318 mg, 0.75 mmol, 1.0 eq.) and $CsN(SiMe_3)_2$ (220 mg, 0.75 mmol, 1.0 eq.) were dissolved in 10 mL of thf and stirred over three days at room temperature. The reaction mixture was filtered and the solvent was evaporated under reduced pressure. The resulting colourless residue was washed *n*-pentane (3*5 mL), which resulted in a white microcrystalline solid. Single crystals suitable for X-ray analysis were obtained after recrystallization from *n*-hexane.

Yield (based on crystals): 276mg (66 %). **Elemental analysis** calcd. (%) for [$C_{56}H_{56}Cs_2N_4P_2$] (1017.95): C 60.44, H 5.03, N 5.05; found: C 60.60, H 5.03, N 5.05.

Path B: 266.2 mg (*R*)-HPEPIA (0.63 mmol ,1.0 eq.) and 100 mg elemental caesium (0.75 mmol, 1.20 eq.) were placed in a Schlenk flask and toluene (20 mL) was added and stirred for 3 h at 70°C. The reaction mixture was allowed to cool to room temperature. After filtration and removal of all volatiles, the off-white residue was recrystallized from *n*-hexane to obtain the product in form of colourless block shaped crystals. **Yield** (based on crystals): 85 mg (20.37%).

^1H NMR (thf-d_8, 300 MHz): δ [ppm] = 7.59-7.53 (m, 8 H, *o*-Ar_{phos}-*CH*), 7.27-7.24 (m, 8 H, *o*-*CH*), 7.16-7.13 (m, 12 H, *m, p*-Ar_{phos}), 7.09-7.04 (m, 8 H, *m*-*CH*), 6.96-6.90 (m, 4 H, *p*-*CH*), 4.07 (dq, $^3J_{HH}$ = 6.4 Hz, $^3J_{PH}$ = 22.7 Hz, 4 H, *CH*), 1.04 (d, $^3J_{HH}$ = 6.4 Hz, 12 H, CH_3).

$^{13}C\{^1H\}$ NMR (thf-d_8, 75 MHz): δ [ppm] = 155.1 (d, $^3J_{PC}$ = 14.9 Hz, Ar-C_q), 143.0 (d, $^1J_{PC}$ = 80.9 Hz, Ar_{phos}-C_q), 133.0 (d, $^2J_{PC}$ = 7.6 Hz, *o*-Ar_{phos}-CH), 128.2 (d, $^4J_{PC}$ = 2.6 Hz, *p*-Ar_{phos}-CH), 128.0 (*m*-CH), 127.6 (*o*-CH), 127.3 (d, $^3J_{PC}$ = 9.4 Hz, *m*-Ar_{phos}-CH), 125.0 (*p*-CH), 55.8 (*C*H), 29.6 (d, $^3J_{PC}$ = 8.1 Hz, CH_3).

^{31}P{^{1}H} NMR (thf-d_8, 121 MHz): δ [ppm] = 12.6.

IR (ATR): $\tilde{\nu}$ [cm^{-1}] = 3494 (vw), 3055 (vw), 3023 (vw), 2961 (vw), 2911 (vw), 2850 (vw), 1596 (vw), 1488 (w), 1449 (w), 1434 (w), 1401 (vw), 1345 (w), 1304 (w), 1272 (w), 1174 (m), 1156 (s), 1131 (m), 1098 (w), 1062 (w), 1026 (w), 1003 (vw), 978 (w), 945 (vw), 911(vw), 843 (m), 820 (m), 807 (m), 779 (vw), 756 (s), 745 (vs), 709 (w), 697 (s), 620 (s), 606 (s), 545 (s), 519 (s), 500 (s), 479 (vw).

Raman (solid state): $\tilde{\nu}$ [cm^{-1}] = 3054 (m), 2962 (vw), 2912 (vw), 2852 (vw), 1599 (w), 1587 (m), 1571 (w), 1453 (vw), 1438 (vw), 1328 (vw), 1307 (vw), 1275 (vw), 1196 (w), 1179 (w), 1145 (w), 1099 (w), 1028 (vs), 1004 (w), 978 (w), 822 (vw), 781 (vw), 764 (vw), 752 (vw), 704 (vw), 676 (vw), 621 (vw),609 (vw), 595 (vw), 554 (w), 521 (vw), 481 (vw).

4.3.6 Synthesis of [{(*R*)-PE^{Dipp}PIA}$_2$Li$_2$] (4)

(*R*)-HPE^{Dipp}PIA (481.0 mg, 1.0 mmol, 1.0 eq.) and LiN(SiMe$_3$)$_2$ (167.0 mg, 1.0 mmol, 1.0 eq.) were dissolved in toluene (30 mL) and stirred at room temperature for 16 h. Removal of the solvent under reduced pressure gave crystalline solid. The resulting white solid was washed with *n*-pentane (5 mL). Single crystals suitable for X-ray analysis were obtained from hot *n*-heptane. The solvent was decanted, and the product was washed with cold *n*-pentane (5 mL).

Yield (based on crystals): 250 mg (51.4 %). **Elemental analysis** calcd. (%) for [$C_{64}H_{72}Li_2N_4P_2$] (973.14): C 78.99, H 7.46, N 5.76; found: C 79.42, H 7.18, N 5.86.

^{1}H NMR (C_6D_6, 300 MHz): δ [ppm] = 7.94-7.89 (m, 4 H, Ar-C*H*), 7.44-7.42 (m, 4 H, Ar-C*H*), 7.11-6.74 (m, 28 H, Ar-C*H*), 4.24-4.08 (m, 4 H, C*H*(CH$_3$)$_2$), 3.12 (br, 2 H, Ph(C*H*)CH$_3$), 1.75 (d, $^{3}J_{HH}$ = 6.64 Hz, 6 H, Ph(CH)C*H*$_3$), 1.38-0.19 (m, 24 H, CH(C*H*$_3$)$_2$).

^{7}Li{^{1}H} NMR (C_6D_6, 117 MHz): δ [ppm] = 2.9

^{13}C{^{1}H} NMR (C_6D_6, 75 MHz): δ [ppm] = 150.4 (d, J_{PC} = 8.7 Hz, Ar-C_q), 143.8 (d, J_{PC} = 3.9 Hz, Ar-C_q), 139.1 (Ar-*C*H), 137.9 (Ar-*C*H), 133.4 (d, J_{PC} = 8.2 Hz, Ar-*C*H), 132.8 (d, J_{PC} = 8.3 Hz, Ar-*C*H), 130.4 (d, J_{PC} = 2.6 Hz, Ar-*C*H), 130.0 (d, J_{PC} = 2.4 Hz, Ar-*C*H), 129.0 (Ar-*C*H), 128.2 (*Ar*-CH), 128.0 (*Ar*-CH), 127.3 (d, J_{PC} = 10.9 Hz, Ar-*C*H), 127.1 (Ar-*C*H), 126.6 (Ar-*C*H), 124.1 (d, J_{PC} = 3.0 Hz, Ar-*C*H), 122.5 (d, J_{PC} = 3.6 Hz, Ar-*C*H), 55.5 (Ph(*C*H)CH$_3$), 28.9 (d, $^{3}J_{PC}$ = 11.5 Hz, Ph(CH)*C*H$_3$), 28.5 (*C*H(CH$_3$)$_2$), 24.4 (CH(*C*H$_3$)$_2$).

$^{31}P\{^1H\}$ NMR (C_6D_6, 121 MHz): δ [ppm] = 28.8

IR (ATR): $\tilde{\nu}$ [cm^{-1}] = 3055 (w), 2958 (m), 2923 (w), 2864 (m), 1494 (w), 1452 (m), 1433 (s), 1378 (m), 1356 (m), 1273 (s), 1254 (m), 1242 (vs), 1231 (w), 1202 (m), 1113 (m), 1085 (m), 1049 (m), 1030 (m), 997 (w), 983 (m), 972 (w), 948 (m), 931 (w), 849 (m), 831 (m), 801 (m), 775 (s), 746 (m), 715 (vs), 696 (m), 667 (m), 607 (m), 598 (vs), 559 (m), 540 (vs), 512 (m), 464 (m).

4.3.7 Synthesis of [{(*R*)-PE^{Dipp}PIA}$_2$Na$_2$] (5)

Following the similar procedure described above for **4** the reaction of (*R*)-HPE^{Dipp}PIA (300.0 mg, 0.62 mmol, 1.0 eq.) and $NaN(SiMe_3)_2$ (114.0 mg, 0.62 mmol, 1.0 eq.) afforded single crystals from hot toluene suitable for X-ray analysis. The solvent was decanted, and the product was washed with cold *n*-pentane (5 mL).

Yield (based on crystals): 200 mg (64.2 %). **Elemental analysis** calcd. (%) for [$C_{64}H_{72}Na_2N_4P_2$] (1005.25): C 76.47, H 7.22, N 5.57; found: C 76.33, H 6.79, N 5.55.

^1H NMR (C_6D_6, 400 MHz): δ [ppm] = 7.67-7.63 (m, 4 H, *o*-Ar$_{phos}$-C*H*), 7.33-7.31 (m, 4 H, Ar-C*H*), 7.07-6.86 (m, 18 H, Ar-CH), 6.65-6.56 (m, 10 H, Ar-C*H*), 4.01 (dq, $^3J_{HH}$ = 6.35 Hz, $^3J_{PH}$ = 25.18 Hz, 2 H, Ph(C*H*)CH$_3$), 3.23 (br, 4H, C*H*(CH$_3$)$_2$), 1.44 (dd, $^3J_{HH}$ = 6.38 Hz, 6 H, Ph(CH)C*H*$_3$), 0.99-0.27 (24 H, CH(C*H*$_3$)$_2$).

$^{13}C\{^1H\}$ NMR (C_6D_6, 100 MHz): δ [ppm] = 152.9 (d, J_{PC} = 10.6 Hz, Ar-C_q), 146.0 (d, J_{PC} = 2.9 Hz, Ar-C_q), 141.4 (Ar-C_q), 140.4 (Ar-C_q), 134.6 (Ar-*C*H), 133.7 (Ar-*C*H), 133.4 (d, J_{PC} = 8.1 Hz, Ar-CH), 131.8 (d, J_{PC} = 8.0 Hz, Ar-*C*H), 129.7 (d, J_{PC} = 2.6 Hz, Ar-*C*H), 129.6 (d, J_{PC} = 2.4 Hz, Ar-*C*H), 128.6 (Ar-*C*H), 128.0 (Ar-*C*H), 127.3 (d, J_{PC} = 10.5 Hz, Ar-*C*H), 127.0 (Ar-*C*H), 124.3 (d, J_{PC} = 3.0 Hz, Ar-*C*H), 121.75 (Ar-*C*H), 55.0 (Ph(*C*H)CH$_3$), 32.3 (*C*H(CH$_3$)$_2$), 29.5 (*C*H(CH$_3$)$_2$), 27.8 (Ph(CH)*C*H$_3$), 23.3 (CH(*C*H$_3$)$_2$), 14.4 (CH(*C*H$_3$)$_2$).

$^{31}P\{^1H\}$ NMR (C_6D_6, 162 MHz): δ [ppm] = 20.8.

IR (ATR): $\tilde{\nu}$ [cm^{-1}] = 3054 (w), 2970 (w), 2950 (w), 2929 (w), 1586 (w), 1493 (s), 1482 (vs), 1450 (s), 1434 (m), 1401 (m), 1378 (m), 1366 (m), 1356 (m), 1314 (m), 1297 (m), 1261 (m), 1202 (m), 1184 (m), 1167 (m), 1152 (m), 1118 (m), 1101 (m), 1087 (s), 1075 (m), 1064 (s), 1047 (m), 1028 (s), 1017 (m), 994 (m), 969 (m), 950 (m), 931 (m), 833 (m), 782 (w), 771 (m), 749 (m), 696 (s), 659 (vs), 610 (s), 600 (m), 573 (vs), 557 (m), 546 (m), 535 (m), 524 (w), 510 (m), 500 (vs), 483 (m), 468 (m), 446 (m), 428 (w).

4.3.8 Synthesis of [{(*R*)-NEPIA}Li(thf)$_2$] (6)

(*R*)-HNEPIA (300.0 mg, 0.58 mmol, 1.0 eq.) and LiN(SiMe$_3$)$_2$ (97.0 mg, 0.58 mmol, 1.0 eq.) were dissolved in thf (30 mL) and stirred at room temperature for 16 h. Removal of the solvent under reduced pressure afforded white solid. The resulting white solid was washed with *n*-pentane (5 mL). Single crystals suitable for X-ray analysis were obtained from concentrated thf solution. The solvent was decanted, and the product was washed with cold *n*-pentane (5 mL).

Yield (based on crystals): 190 mg (48.5 %). **Elemental analysis** calcd. (%) for [$C_{44}H_{48}LiN_2PO_2$] = {**6**·(thf)$_2$} (674.8): C 78.32, H 7.17, N 4.15; found: C 77.94, H 6.58, N 4.48.

^{1}H NMR (C_6D_6, 400 MHz): δ [ppm] = 7.98 (d, J_{PH} = 8.5 Hz, 2 H, Ar-C*H*), 7.80-7.66 (m, 8 H, Ar-C*H*), 7.52 (d, J_{PH} = 8.14 Hz, 2 H, Ar-C*H*), 7.23-7.19 (m, 2 H, Ar-C*H*), 7.13-7.06 (m, 4 H, Ar-C*H*), 6.94-6.90 (m, 6 H, Ar-C*H*), 5.17 (dq, $^3J_{HH}$ = 6.2 Hz, $^3J_{PH}$ = 19.79 Hz, 2 H, naph(C*H*)CH$_3$), 1.17 (d, $^3J_{HH}$ = 6.32 Hz, 6 H, naph(CH)C*H*$_3$).

^{7}Li{^{1}H} NMR (C_6D_6, 117 MHz): δ [ppm] = 2.7

^{13}C{^{1}H} NMR (C_6D_6, 75 MHz): δ [ppm] = 147.0 (d, J_{PC} = 12.7 Hz, Ar-*C*$_q$), 136.9 (Ar$_{phos}$-*C*$_q$), 136.1 (Ar$_{phos}$-*C*$_q$), 134.9 (Ar$_{phos}$-*C*$_q$), 133.0 (d, J_{PC} = 8.4 Hz, Ar-*C*H), 131.0 (Ar-*C*H), 130.3 (Ar-*C*H), 129.3 (Ar-*C*H), 127.9 (*Ar*-CH), 127.1 (*Ar*-CH), 126.4 (Ar-*C*H), 125.8 (*Ar*-CH), 124.2 (*Ar*-CH), 122.4 (*Ar*-CH), 48.5 (naph(*C*H)CH$_3$), 28.9 (d, $^3J_{PC}$ = 6.4 Hz, naph(CH)*C*H$_3$).

^{31}P{^{1}H} NMR (C_6D_6, 121 MHz): δ [ppm] = 30.4

IR (ATR): $\tilde{\nu}$ [cm^{-1}] = 3047 (vw), 2975 (w), 2949 (vw), 2912 (w), 2874 (w), 2851 (vw), 1594 (w), 1509 (w), 1433 (w), 1392 (m), 1357 (w), 1309 (w), 1255 (w), 1235 (s), 1207 (vs), 1179 (s), 1153 (s), 1104 (s), 1050 (m), 1026 (m), 977 (m), 963 (m), 893 (m), 852 (m), 820 (s), 795 (vs), 775 (vs), 749 (s), 721 (s), 700 (s), 670 (w), 648 (w), 622 (vw), 574 (vs), 547 (vs), 512 (vs), 495 (vs), 446 (m), 433 (w).

4.3.9 Synthesis of [{(*R*)-NEPIA}$_2$Na$_2$] (7)

Following the similar procedure described above for **6** the reaction of [(*R*)-HNEPIA] (300.0 mg, 0.58 mmol, 1.0 eq.) and NaN(SiMe$_3$)$_2$ (105.0 mg, 0.58 mmol, 1.0 eq.) afforded single crystals from hot toluene suitable for X-ray analysis. The solvent was decanted, and the product was washed with cold *n*-pentane (5 mL).

Yield (based on crystals): 160 mg (50 %). **Elemental analysis** calcd. (%) for [$C_{72}H_{64}N_4P_2Na_2$] (1093.26): C 79.10, H 5.90, N 5.12; found: C C 78.37, H 5.75, N 5.10.

^{1}H NMR (C_6D_6, 400 MHz): δ [ppm] = 7.79 (m, 4 H, *o*-Ar$_{phos}$-C*H*), 7.52-7.50 (br, 12 H, Ar-C*H*), 7.42-7.40 (m, 4 H, Ar-CH), 7.32-7.23 (m, 16 H, Ar-C*H*), 7.06 (br, 4 H, Ar-C*H*), 6.94-6.79 (m, 8 H, Ar-C*H*), 4.61 (dq, $^3J_{HH}$ = 6.37 Hz, $^3J_{PH}$ = 19.63 Hz, 4 H, naph(C*H*)C*H*$_3$), 1.26 (d, $^3J_{HH}$ = 6.37 Hz, 12 H, naph(CH)C*H*$_3$).

^{13}C{^{1}H} NMR (C_6D_6, 100 MHz): δ [ppm] = 148.6 (d, J_{PC} = 16.6 Hz, Ar-C_q), 138.1 (d, J_{PC} = 72.3 Hz, Ar-C_q), 134.4 (Ar-C_q), 132.6 (d, J_{PC} = 8.0 Hz, Ar-*C*), 130.6 (Ar-*C*H), 129.6 (Ar-*C*H), 129.1 (Ar-*C*H), 128.2 (Ar-*C*H), 127.9 (Ar-*C*H), 126.6 (Ar-*C*H), 125.9 (Ar-*C*H), 125.5 (Ar-*C*H), 123.4 (Ar-*C*H), 122.9 (Ar-*C*H), 48.4 (naph(*C*H)CH$_3$), 27.5 (naph(CH)*C*H$_3$).

^{31}P{^{1}H} NMR (C_6D_6, 162 MHz): δ [ppm] = 24.8.

IR (ATR): $\tilde{\nu}$ [cm^{-1}] = 3049 (w), 2975 (w), 2956 (w), 2916 (vw), 2857 (vw), 1594 (m), 1509 (m), 1479 (vw), 1452 (vw), 1432 (m), 1394 (w), 1361 (w), 1296 (m), 1255 (vw), 1232 (m), 1175 (m), 1158 (vs), 1127 (m), 1099 (m), 1027 (m), 998 (m), 971 (vw), 864 (m), 848 (vs), 811 (vs), 794 (vs), 775 (vs), 747 (vs), 724 (vs), 699 (vs), 647 (vs), 613 (m), 574 (m), 537 (vs), 516 (vs), 497 (vs), 469 (m), 430 (m).

4.3.10 Synthesis of [{(*R*)-NEPIA}$_2$K$_2$] (8)

Following the similar procedure described above for **7** the reaction of [(*R*)-HNEPIA] (525.0 mg, 1.0 mmol, 1.0 eq.) and KN(SiMe$_3$)$_2$ (199.0 mg, 1.0 mmol, 1.0 eq.) afforded single crystals from hot toluene suitable for X-ray analysis. The solvent was decanted, and the product was washed with cold *n*-pentane (5 mL).

Yield (based on crystals): 400 mg (71 %). **Elemental analysis** calcd. (%) for [$C_{72}H_{64}N_4P_2K_2$] (1125.48): C 76.84, H 5.73, N 4.98; found: C 77.06, H 5.49, N 4.86.

^{1}H NMR (C_6D_6, 400 MHz): δ [ppm] = 7.80-7.73 (m, 12 H, Ar-C*H*), 7.61-7.50 (m, 12 H, Ar-C*H*), 7.35-7.33 (m, 12 H, Ar-CH), 7.13-7.10 (m, 8 H, Ar-C*H*), 6.96-6.92 (m, 4 H, Ar-C*H*), 4.68 (dq, $^3J_{HH}$ = 6.38 Hz, $^3J_{PH}$ = 20.57 Hz, 4 H, naph(C*H*)C*H*$_3$), 1.58 (d, $^3J_{HH}$ = 6.41 Hz, 12 H, naph(CH)C*H*$_3$).

^{13}C{^{1}H} NMR (C_6D_6, 100 MHz): δ [ppm] = 148.8 (d, J_{PC} = 18.9 Hz, Ar-C_q), 140.2 (Ar-C_q), 139.4 (Ar-C_q), 134.3 (Ar-C_q), 132.5 (d, J_{PC} = 7.5 Hz, Ar-*C*), 131.1 (Ar-*C*H), 129.0 (Ar-*C*H), 128.8 (Ar-*C*H),

127.7 (d, J_{PC} = 9.5 Hz, Ar-*C*H), 126.7 (Ar-*C*H), 125.8 (Ar-*C*H), 125.4 (d, J_{PC} = 10.4 Hz, Ar-*C*H), 124.4 (Ar-*C*H), 123.2 (Ar-*C*H), 49.8 (naph(*C*H)CH$_3$), 27.4 (naph(CH)*C*H$_3$).

^{31}P{^{1}H} NMR (C_6D_6, 162 MHz): δ [ppm] = 18.5.

IR (ATR): $\tilde{\nu}$ [cm^{-1}] = 3045 (w), 2950 (m), 2913 (w), 1594 (m), 1507 (m), 1432 (s), 1394 (w), 1360 (m), 1297 (w), 1156 (vs), 1117 (s), 1095 (s), 1082 (m), 1065 (m), 1026 (m), 1010 (w), 966 (w), 950 (m), 845 (m), 810 (s), 792 (s), 777 (s), 748 (s), 700 (vs), 649 (m), 614 (m), 574 (vs), 539 (vs), 514 (vs), 498 (w), 469 (vs), 451 (m), 430 (w).

4.4 Synthesis and analytical data for alkaline earth metal complexes of chiral iminophosphonamide

4.4.1 Synthesis of [{(*R*)-PEPIA}$_2$Ca] (9)

[Ca{N(SiMe$_3$)$_2$}$_2$(thf)$_2$] (182.0 mg, 0.36mmol, 1.0 eq.) and (*R*)-HPEPIA (300.0 mg, 0.71 mmol, 2.0 eq.) were dissolved in toluene (30 mL) and stirred at room temperature for 16 h. After removal of the solvent under reduced pressure a white solid was obtained. The resulting white solid was washed with *n*-pentane (5 mL). Single crystals suitable for X-ray analysis were obtained from hot *n*-hexane. The solvent was decanted, and the product was washed with cold *n*-pentane (5 mL).

Yield (based on crystals): 190 mg (59.5 %). **Elemental analysis** calcd. (%) for [$C_{56}H_{56}CaN_4P_2$] (887.09): C 75.82, H 6.36, N 6.32; found: C 75.56, H 6.20, N 6.46.

^{1}H NMR (C_6D_6, 400 MHz): δ [ppm] = 7.71-7.66 (m, 8 H, *o*-Ar$_{phos}$-C*H*), 7.21-7.19 (m, 16 H, *o*-Ar-C*H* & *m*-Ar$_{phos}$-C*H*), 7.18-7.17 (m, 4 H, Ar-C*H*), 7.15-7.13 (m, 6 H, Ar-C*H*), 7.11-7.06 (m, 6 H, Ar-C*H*), 4.04 (dq, $^3J_{HH}$ = 6.5 Hz, $^3J_{PH}$ = 19.6 Hz, 4 H, Ph(C*H*)CH$_3$), 1.38 (d, $^3J_{HH}$ = 6.5 Hz, 12 H, Ph(CH)C*H*$_3$).

^{13}C{^{1}H} NMR (C_6D_6, 100 MHz): δ [ppm] = 152.0 (d, $^3J_{PC}$ = 14.0 Hz, Ar-*C*$_q$), 135.9 (d, $^1J_{PC}$ = 85.4 Hz, Ar$_{phos}$-*C*$_q$), 132.7 (d, $^2J_{PC}$ = 8.8 Hz, *o*-Ar$_{phos}$-CH), 130.3 (d, $^4J_{PC}$ = 2.7 Hz, Ar-CH), 129.3 (Ar-CH), 127.9 (Ar-CH), 126.1 (Ar-*C*H), 125.9 (*Ar*-CH), 54.1 (Ph(*C*H)CH$_3$), 28.4 (d, $^3J_{PC}$ = 6.7 Hz, Ph(CH)*C*H$_3$).

^{31}P{^{1}H} NMR (C_6D_6, 162 MHz): δ [ppm] = 24.5

IR (ATR): $\tilde{\nu}$ [cm^{-1}] = 3058 (vw), 3022 (vw), 2957 (vw), 2913 (vw), 2835 (vw), 1598 (vw), 1489 (m), 1479 (m), 1452 (m), 1433 (m), 1401 (m), 1361 (m), 1346 (m), 1309 (m), 1273 (w), 1236 (s), 1204 (s), 1176 (m), 1167 (m), 1156 (m), 1149 (m), 1118 (m), 1102 (w), 1087 (m), 1060 (m), 1025 (m), 1000 (s), 984 (m), 973 (m), 965 (s), 945 (vs), 911 (m), 859 (m), 837 (m), 822 (m), 780 (s), 772 (s), 761 (m), 749 (m), 719 (s), 714 (s), 695 (s), 622 (m), 608 (m), 597 (m), 574 (m), 565 (m), 510 (s), 463 (s), 446 (m), 434 (vw).

4.4.2 Synthesis of [{(*R*)-PE^{Dipp}PIA}{N(SiMe$_3$)$_2$}Mg(thf)] (10)

Following the similar procedure described above for **9**, the reaction of [Mg{N(SiMe$_3$)$_2$}$_2$(thf)$_2$] (158.0 mg, 0.46 mmol, 1.0 eq.) and (*R*)-HPE^{Dipp}PIA (221.1 mg, 0.46 mmol, 1.0 eq.) in thf (20 mL) afforded single crystals from *n*-pentane suitable for X-ray analysis. The solvent was decanted, and the product was washed with cold *n*-pentane (5 mL).

Yield (based on crystals): 220.0 mg (65.0 %). **Elemental analysis** calcd. (%) for [$C_{42}H_{62}MgN_3POSi_2$] (736.43): C 68.50, H 8.49, N 5.71; found: C 68.53, H 8.34 N 5.55.

^{1}H NMR (C_6D_6, 400 MHz): δ [ppm] = 7.51-7.46 (m, 2 H, *o*-Ar$_{phos}$-C*H*), 7.18 (m, 1 H, Ar-C*H*), 7.09-7.04 (m, 5 H, Ar-C*H*), 7.00-6.88 (m, 7 H, Ar-C*H*), 6.81-6.78 (m, 1 H, Ar-C*H*), 6.71-6.66 (m, 2 H, Ar-C*H*), 3.92-3.82 (m, 5 H, thf-*H* & Ph(C*H*)C*H*$_3$), 3.44-3.37 (m, 2 H, C*H*(CH$_3$)$_2$), 1.46 (d, $^3J_{HH}$ = 6.7 Hz, 3 H, Ph(CH)C*H*$_3$), 1.35-1.32 (m, 4H, thf-*H*), 1.16-0.52 (m, 12 H, CH(C*H*$_3$)$_2$), 0.37 (s, 18 H, N(SiMe$_3$)$_2$).

^{13}C{^{1}H} NMR (C_6D_6, 100 MHz): δ [ppm] = 149.5 (d, J_{PC} = 5.5 Hz, Ar-C_q), 146.3 (d, J_{PC} = 6.1 Hz, Ar-C_q), 141.1 (d, J_{PC} = 4.4 Hz, Ar-C_q), 134.3 (Ar-C_q), 133.3 (d, J_{PC} = 8.9 Hz, Ar-CH), 133.0 (d, J_{PC} = 8.3, Ar-CH), 132.4 (Ar-CH), 131.5 (Ar-CH), 130.9 (d, J_{PC} = 2.6 Hz, Ar-CH), 130.6 (d, J_{PC} = 2.7 Hz, Ar-CH), 128.4 (Ar-CH), 127.5 (Ar-CH), 127.4 (Ar-CH), 126.5 (Ar-CH), 123.9 (d, J_{PC} = 3.8 Hz, Ar-CH), 123.5 (Ar-CH), 69.8 (^{thf}CH$_2$O), 55.0 (Ph(*C*H)CH$_3$), 32.3 (*C*H(CH$_3$)$_2$), 29.2 (d, J_{PC} = 14.4 Hz, Ph*C*H(CH$_3$)), 25.2 (CH(*C*H$_3$)$_2$), 23.1 (^{thf}CH$_2$), 6.6 (N(Si*Me*$_3$)$_2$).

^{31}P{^{1}H} NMR (C_6D_6, 162 MHz): δ [ppm] = 31.9

IR (ATR): $\tilde{\nu}$ [cm^{-1}] = 3057 (w), 2964 (s), 2925 (m), 2865 (m), 1589 (vw), 1487 (vs), 1453 (m), 1401 (m), 1375 (vw), 1362 (m), 1318 (s), 1299 (m), 1252 (m), 1236 (m), 1198 (m), 1138 (m), 1107 (m), 1069 (m), 1048 (m), 1027 (m), 1003 (m), 973 (m), 950 (m), 929 (s), 889 (vs), 859 (m), 839 (m), 820 (s), 787 (vs), 746 (s), 718 (m), 696 (m), 668 (m), 616 (m), 593 (m), 571 (m), 542 (s), 516 (vs), 496 (vs), 454 (m), 417 (m).

4.4.3 Synthesis of [{(*R*)-PE^{Dipp}PIA}$_2$Mg] (11)

Following the similar procedure described above for **9**, the reaction of (*R*)-HPE^{Dipp}PIA (300 mg, 0.23 mmol, 2.0 eq.) and [Mg{N(SiMe$_3$)$_2$}$_2$(thf)$_2$] (54.0 mg, 0.12 mmol, 1.0 eq.) in 30 mL afforded single crystals from *n*-pentane suitable for X-ray analysis. The solvent was decanted, and the product was washed with cold *n*-pentane (5 mL).

Yield (based on crystals): 50.0 mg (11.0 %). **Elemental analysis** calcd. (%) for [$C_{64}H_{72}MgN_4P_2$] (983.53): C 78.16, H 7.38, N 5.70; found: C 78.54, H 7.63, N 5.70.

^{1}H NMR (C_6D_6, 400 MHz): δ [ppm] = 7.87-7.25 (m, 12 H, Ar-C*H*), 7.13-6.95 (m, 20 H, Ar-C*H*), 6.83 (br, 4 H, Ar-CH), 4.03-3.88 (m, 4 H, C*H*(CH$_3$)$_2$), 3.26 (br, 2 H, Ph(C*H*)CH$_3$), 2.02-0.36 (m, 30 H, Ph(C*H*)CH$_3$ & CH(C*H*$_3$)$_2$).

^{13}C{^{1}H} NMR (C_6D_6, 100 MHz): δ [ppm] = 146.3 (Ar-C_q), 146.2 (Ar-C_q), 141.9 (d, J_{PC} = 4.2 Hz, Ar-C_q), 133.5 (Ar-CH), 132.1 (Ar-CH), 130.9 (Ar-CH), 130.4 (Ar-CH), 129.3 (Ar-CH), 128.7 (Ar-CH), 128.6 (Ar-CH), 127.9 (Ar-CH), 127.3 (Ar-CH), 126.9 (Ar-CH), 125.7 (Ar-CH), 124.3 (Ar-CH), 123.2 (Ar-CH), 55.8 (Ph(*C*H)CH$_3$), 26.2 (*C*H(CH$_3$)$_2$), 25.3 (Ph(CH)*C*H$_3$), 25.1 (CH(*C*H$_3$)$_2$).

^{31}P{^{1}H} NMR (C_6D_6, 162 MHz): δ [ppm] = 28.6

IR (ATR): $\tilde{\nu}$ [cm^{-1}] = 3057 (w), 2964 (s), 2925 (m), 2865 (m), 1589 (vw), 1487 (vs), 1453 (m), 1401 (m), 1375 (vw), 1362 (m), 1318 (s), 1299 (m), 1252 (m), 1236 (m), 1198 (m), 1138 (m), 1107 (m), 1069 (m), 1048 (m), 1027 (m), 1003 (m), 973 (m), 950 (m), 929 (s), 889 (vs), 859 (m), 839 (m), 820 (s), 787 (vs), 746 (s), 718 (m), 696 (m), 668 (m), 616 (m), 593 (m), 571 (m), 542 (s), 516 (vs), 496 (vs), 454 (m), 417 (m).

4.4.4 Synthesis of [{(*R*)-PE^{Dipp}PIA}$_2$Ca] (12)

Following the similar procedure described above for **11** the reaction of [Ca{N(SiMe$_3$)$_2$}$_2$(thf)$_2$] (157.0 mg, 0.31mmol, 1.0 eq.) and (*R*)-HPE^{Dipp}PIA (300 mg, 0.62 mmol, 2.0 eq.) afforded single crystals from *n*-pentane suitable for X-ray analysis. The solvent was decanted, and the product was washed with cold *n*-pentane (5 mL).

Yield (based on crystals): 220.0 mg (71.0 %). **Elemental analysis** calcd. (%) for [$C_{64}H_{72}CaN_4P_2$] (999.33): C 76.92, H 7.26, N 5.61; found: C 77.13, H 7.69, N 5.82.

^{1}H NMR (C_6D_6, 400 MHz): δ [ppm] = 7.50-7.45 (m, 4 H, *o*-Ar$_{phos}$-C*H*), 7.27-7.25 (m, 4 H, Ar-C*H*), 7.14-7.09 (m, 6 H, Ar-CH), 6.99-6.94 (m, 10 H, Ar-C*H*), 6.83-6.74 (m, 12 H, Ar-C*H*), 3.67 (dq, $^3J_{HH}$ = 6.5 Hz, $^3J_{PH}$ = 24.87 Hz, 2 H, Ph(C*H*)C*H*$_3$), 3.07 (br, 4 H, C*H*(CH$_3$)$_2$), 1.65 (d, $^3J_{HH}$ = 6.5 Hz, 6 H, Ph(CH)C*H*$_3$), 1.06-0.38 (m, 24 H, CH(C*H*$_3$)$_2$).

^{13}C{^{1}H} NMR (C_6D_6, 100 MHz): δ [ppm] = 152.5 (d, $^3J_{PC}$ = 8.9 Hz, Ar-C_q), 144.6 (Ar-C_q), 143.0 (d, J_{PC} = 4.3 Hz, Ar-C_q), 135.8 (Ar-*C*), 134.9 (d, $^2J_{PC}$ = 8.6 Hz, Ar-CH), 133.9 (Ar-CH), 133.5 (d, J_{PC} = 8.9 Hz, Ar-CH), 132.4 (d, J_{PC} = 8.1 Hz, Ar-CH), 130.4 (d, J_{PC} = 2.5 Hz, Ar-CH), 130.2 (d, J_{PC} = 2.6 Hz, Ar-CH), 129.5 (Ar-CH), 127.7 (Ar-CH), 127.5 (d, J_{PC} = 10.9 Hz, Ar-CH), 126.5 (Ar-CH), 123.3 (Ar-CH), 122.02 (d, J_{PC} = 4.8 Hz, Ar-CH), 55.8 (Ph(CH)CH$_3$), 29.6 (d, $^3J_{PC}$ = 10.9 Hz, Ph(CH)*C*H$_3$), 28.5 (*C*H(CH$_3$)$_2$), 24.1 (br, CH(*C*H$_3$)$_2$).

^{31}P{^{1}H} NMR (C_6D_6, 162 MHz): δ [ppm] = 22.6.

IR (ATR): $\tilde{\nu}$ [cm^{-1}] = 3055 (vw), 2964 (m), 2954 (w), 2919 (w), 2863 (w), 1588 (w), 1492 (m), 1452 (s), 1432 (w), 1380 (vw), 1364 (vw), 1327 (vw), 1308 (m), 1255 (m), 1204 (vs), 1179 (s), 1146 (w), 1111 (w), 1099 (w), 1067 (w), 1047 (m), 1026 (m), 1004 (vw), 992 (vw), 973 (vw), 950 (vw), 931 (vw), 837 (vs), 787 (m), 775 (m), 756 (m), 743 (m), 715 (m), 696 (vs), 662 (m), 609 (m), 583 (m), 564 (s), 524 (s), 514 (m), 490 (m), 477 (m), 451 (m).

4.4.5 Synthesis of [{(*R*)-PE^{Dipp}PIA}$_2$Yb] (13)

Toluene (30 mL) was added to the mixture of [Yb{N(SiMe$_3$)$_2$}(thf)$_2$] (133 mg, 0.21 mmol, 1.0 eq.) and (*R*)-HPE^{Dipp}PIA (200 mg, 0.42 mmol, 2.0 eq.). The resulting reaction mixture was stirred for 16 h at room temperature. After removing the solvent under reduced pressure resulted into red powder. Single crystals suitable for X-ray crystallography were afforded from concentrated *n*-pentane solution at -30°C. The solvent was decanted and crystals were with cold *n*-pentane.

Yield (based on crystals): 150 mg (63 %). **Elemental analysis** calcd. (%) for [$C_{64}H_{72}YbN_4P_2$] (1132.27): C 67.89, H 6.41, N 4.95; found: C 68.67, H 6.64, N 5.07.

^{1}H NMR (C_6D_6, 400 MHz): δ [ppm] = 7.80-7.77 (m, 4 H, Ar-C*H*), 7.49-7.47 (m, 4 H, Ar-C*H*), 7.34-7.30 (m, 4 H, Ar-C*H*), 7.20-7.13 (m, 12 H, Ar-C*H*), 7.02-6.90 (m, 12 H, Ar-C*H*), 4.01 (dq, 2 H, $^3J_{HH}$ = 6.40 Hz, $^3J_{PH}$ = 24.0, Ph(C*H*)CH$_3$), 3.37 (br, 4 H, C*H*(CH$_3$)$_2$), 1.84 (d, 6 H, $^3J_{HH}$ = 6.40 Hz, Ph(CH)C*H*$_3$), 1.19-0.70 (m, 24 H, CH(C*H*$_3$)$_2$).

^{13}C{^{1}H} NMR (C_6D_6, 75 MHz): δ [ppm] = 152.2 (d, J_{PC} = 9.3 Hz, Ar-C_q), 144.5 (d, J_{PC} = 6.4 Hz, Ar-C_q), 143.6 (d, J_{PC} = 3.7 Hz, Ar-C_q), 135.9 (Ar-*C*H), 134.9 (d, J_{PC} = 89.4 Hz, Ar-*C*H), 132.9 (d, J_{PC} = 8.8 Hz, Ar-*C*H), 132.5 (d, J_{PC} = 8.3 Hz, Ar-*C*H), 130.4 (d, J_{PC} = 2.6 Hz, Ar-*C*H), 130.1 (d, J_{PC} = 2.7 Hz, Ar-*C*H), 129.3 (Ar-*C*H), 127.5 (d, J_{PC} = 10.9 Hz, Ar-*C*H), 126.7 (Ar-*C*H), 126.4 (Ar-*C*H), 126.2 (Ar-*C*H), 123.4 (d, J_{PC} = 3.4 Hz, Ar-*C*H), 121.9 (d, J_{PC} = 4.1 Hz, Ar-*C*H), 55.6 (Ph(*C*H)CH_3), 34.5 (*C*H(CH_3)$_2$), 29.3 (d, J_{PC} = 10.2 Hz, Ph(CH)*C*H$_3$), 28.5 (CH(*C*H$_3$)$_2$), 23.5 (CH(*C*H$_3$)$_2$), 22.7 (CH(*C*H$_3$)$_2$), 14.3 (CH(*C*H$_3$)$_2$).

^{31}P{^{1}H} NMR (C_6D_6, 121 MHz): δ [ppm] = 20.6

IR (ATR): $\tilde{\nu}$ [cm^{-1}] = 3058 (w), 3024 (vw), 2967 (w), 2856 (w), 1600 (w), 1491 (m), 1452 (m), 1436 (m), 1372 (m), 1310 (m), 1274 (m), 1202 (s), 1154 (vs), 1122 (m), 1104 (m), 1098 (m), 1070 (m), 1048 (m), 1028 (m), 1003 (w), 985 (vw), 909 (vs), 835 (m), 776 (s), 742 (s), 715 (vs), 694 (vs), 634 (m), 616 (m), 580 (m), 561 (m), 524 (vs), 517 (vs), 465 (vw), 436 (vw), 433 (vw).

4.4.6 Synthesis of [{(*R*)-NEPIA}$_2$Ca] (14)

Following the similar procedure described above for **11** the reaction of [Ca{N(SiMe$_3$)$_2$}$_2$(thf)$_2$] (96.0 mg, 0.19 mmol, 1.0 eq.) and (*R*)-HNEPIA (200.0 mg, 0.38 mmol, 2.0 eq.) afforded single crystals from *n*-pentane suitable for X-ray analysis. The solvent was decanted, and the product was washed with cold *n*-pentane (5 mL).

Yield (based on crystals): 120.0 mg (58.1 %). **Elemental analysis** calcd. (%) for [$C_{72}H_{64}CaN_4P_2$] (1087.33): C 79.53, H 5.93, N 5.15; found: C 79.72, H 6.28, N 5.18.

^{1}H NMR (C_6D_6, 400 MHz): δ [ppm] = 7.77-7.70 (m, 8 H, *o*-Ar$_{phos}$-C*H*), 7.65-7.60 (m, 12 H, Ar$_{phos}$-C*H*), 7.48-7.46 (m, 4 H, Ar-C*H*), 7.14-7.10 (m, 4 H, Ar-C*H*), 7.06-6.99 (m, 20 H, Ar-C*H*), 4.82 (dq, $^{3}J_{HH}$ = 6.4 Hz, $^{3}J_{PH}$ = 19.02 Hz, 4 H, naph(C*H*)CH_3), 1.39 (d, $^{3}J_{HH}$ = 6.44 Hz, 12 H, naph(CH)C*H*$_3$).

^{13}C{^{1}H} NMR (C_6D_6, 100 MHz): δ [ppm] = 147.4 (d, $^{3}J_{PC}$ = 12.67 Hz, Ar-C_q), 135.8 (Ar-*C*), 134.9 (Ar-*C*), 134.6 (Ar-*C*H), 132.7 (d, J_{PC} = 8.7 Hz, Ar-*C*H), 130.9 (Ar-*C*H), 130.2 (d, J_{PC} = 2.5 Hz, Ar-*C*H), 129.5 (Ar-*C*H), 126.7 (Ar-*C*H), 126.5 (Ar-*C*), 126.0 (Ar-*C*), 125.3 (Ar-*C*), 123.3 (Ar-*C*), 123.2 (Ar-*C*), 49.7 (naph(*C*H)CH_3), 28.9 (d, $^{3}J_{PC}$ = 8.3 Hz, naph(CH)*C*H$_3$).

^{31}P{^{1}H} NMR (C_6D_6, 162 MHz): δ [ppm] = 27.3

IR (ATR): $\tilde{\nu}$ [cm^{-1}] = 3052 (vw), 2959 (vw), 2917 (w), 2859 (w), 1595 (w), 1508 (w), 1481 (w), 1456 (m), 1434 (w), 1394 (w), 1364 (w), 1323 (w), 1310 (w), 1300 (w), 1256 (vw), 1235 (vw), 1172 (w), 1144 (w), 1133 (w), 1082 (vs), 1027 (m), 1014 (m), 1001 (w), 970 (m), 926 (m), 908 (m), 849 (s), 835 (s), 821 (s), 791 (m), 775 (m), 748 (m), 730 (m), 712 (vs), 697 (w), 645 (m), 628 (vw), 617 (m), 575 (m), 560 (m), 543 (vs), 532 (vs), 517 (m), 499 (m), 488 (vw), 432 (m), 420 (vw).

4.5 Synthesis of copper and zinc complexes of iminophosphonamides

4.5.1 Synthesis of [{(*R*)-PE^{Dipp}PIA}Cu]$_2$ (**15**)

CuCl (38.0 mg, 0.38mmol, 1.0 eq.), (*R*)-HPE^{Dipp}PIA (183.0 mg, 0.38 mmol, 1.0 eq.) and KN(SiMe$_3$)$_2$ (76.0 mg, 0.38 mmol, 1.0 eq.) were dissolved in thf (30 mL) in a one pot and stirred at room temperature for 16 h. After extracting with toluene and filtration afforded white solid. The resulting white solid was washed with *n*-pentane (5 mL). Single crystals suitable for X-ray analysis were obtained from hot toluene. The solvent was decanted, and the product was washed with cold *n*-pentane (5 mL).

Yield (based on crystals): 150 mg (34 %). **Elemental analysis** calcd. (%) for [$C_{71}H_{80}Cu_2N_4P_2$] {**15**·toluene} (1178.46): C 72.36, H 6.84, N 4.75; found: C 72.99, H 7.07, N 4.99.

^{1}H NMR (C_6D_6, 400 MHz): δ [ppm] = 7.97 (br, 4 H, Ar-C*H*), 7.46-7.44 (m, 4 H, Ar-C*H*), 7.32-7.28 (m, 4 H, Ar-C*H*), 7.14-7.09 (m, 9 H, Ar-C*H*), 7.01-6.76 (m, 15 H, Ar-C*H*), 4.18-4.11 (m, 2 H, Ph(C*H*)CH$_3$), 3.90-3.74 (m, 4 H, C*H*(CH$_3$)$_2$), 1.47 (d, $^3J_{HH}$ = 6.42 Hz, 6 H, CH(C*H*$_3$)$_2$), 1.44 (d, $^3J_{HH}$ = 6.76 Hz, 6 H, CH(C*H*$_3$)$_2$), 1.31 (d, $^3J_{HH}$ = 6.70 Hz, 6 H, Ph(CH)C*H*$_3$), 0.65 (d, $^3J_{HH}$ = 6.78 Hz, 6 H, CH(C*H*$_3$)$_2$), 0.43 (d, $^3J_{HH}$ = 6.78 Hz, 6 H, CH(C*H*$_3$)$_2$).

^{13}C{^{1}H} NMR (C_6D_6, 100 MHz): δ [ppm] = 148.5 (d, $^3J_{PC}$ = 8.4 Hz, Ar-C_q), 147.5 (d, $^1J_{PC}$ = 5.0 Hz, Ar$_{phos}$-C_q), 146.7 (d, J_{PC} = 5.3 Hz, Ar-*C*), 141.2 (Ar-*C*), 133.7 (d, $^3J_{PC}$ = 9.2 Hz, Ar-*C*H), 132.9 (d, $^3J_{PC}$ = 8.6 Hz, Ar-*C*H), 130.6 (Ar-*C*H), 128.2 (Ar-*C*H), 128.0 (d, J_{PC} = 2.7 Hz, Ar-*C*H), 127.5 (Ar-*C*H), 127.4 (Ar-*C*H), 127.2 (Ar-*C*H), 126.4 (Ar-*C*H), 124.2 (Ar-*C*H), 124.02 (Ar-*C*H), 123.93 (Ar-*C*H), 56.2 (d, $^2J_{PC}$ = 4.8 Hz, Ph(*C*H)CH$_3$), 30.9 (*C*H(CH$_3$)$_2$), 28.5 (d, $^3J_{PC}$ = 23.1 Hz, Ph(CH)*C*H$_3$), 24.2 (d, $^3J_{PC}$ = 26.0 Hz, Ph(CH)*C*H$_3$), 23.3 (CH(*C*H$_3$)$_2$), 22.4 (CH(*C*H$_3$)$_2$).

^{31}P{^{1}H} NMR (C_6D_6, 121 MHz): δ [ppm] = 38.0.

IR (ATR): $\tilde{\nu}$ [cm^{-1}] = 3075 (vw), 3051 (vw), 2966 (m), 2947 (w), 2864 (w), 1587 (vw), 1489 (w), 1461 (w), 1450 (m), 1433 (s), 1381 (m), 1363 (m), 1309 (m), 1270 (m), 1191 (s), 1174 (s), 1119 (s), 1107 (s), 1090 (m), 1067 (s), 1052 (m), 1027 (s), 1005 (s), 997 (m), 975 (s), 931 (m), 887 (s), 833 (m), 796 (m), 763 (m), 754 (m), 746 (s), 730 (m), 716 (m), 692 (vs), 623 (w), 605 (m), 577 (s), 545 (m), 529 (vs), 512 (m), 503 (vs), 494 (m), 464 (m), 450 (m), 414 (vw).

4.5.2 Synthesis of [{(*R*)-PE^{Dipp}PIA}$_2$Zn] (16)

$ZnPh_2$ (220.0 mg, 1.0 mmol, 1.0 eq.) and (*R*)-HPE^{Dipp}PIA (481.0 mg, 2.0 mmol, 2.0 eq.) was dissolved in toluene (30mL) and stirred afforded for 16 h. The solvent was removed under reduced pressure and product was washed with *n*-Pentane. The single crystals suitable for X-ray analysis were obtained from *n*-pentane. The solvent was decanted, and the product was washed with cold *n*-pentane (5 mL).

Yield (based on crystals): 410 mg (40 %). **Elemental analysis** calcd. (%) for [$C_{64}H_{72}ZnN_4P_2$] (1024.62): C 75.02, H 7.08, N 5.47; found: C 74.87, H 6.68, N 5.32.

^{1}H NMR (C_6D_6, 400 MHz): δ [ppm] = 7.68-7.66 (m, 4 H, *o*-Ar$_{phos}$-C*H*), 7.40-7.35 (m, 4 H, Ar-C*H*), 7.14-7.07 (m, 10 H, Ar-CH), 6.94-6.84 (m, 12 H, Ar-C*H*), 6.76-6.72 (m, 2 H, Ar-C*H*), 6.65-6.61 (m, 4 H, Ar-C*H*), 3.79 (dq, $^3J_{HH}$ = 6.53 Hz, $^3J_{PH}$ = 18.57 Hz, 2 H, Ph(C*H*)CH_3), 3.42 (br, 4 H, C*H*(CH_3)$_2$), 1.37 (d, $^3J_{HH}$ = 6.52 Hz, 6 H, Ph(CH)CH_3), 0.75 (br, 24 H, CH(CH_3)$_2$).

^{13}C{^{1}H} NMR (C_6D_6, 100 MHz): δ [ppm] = 149.2 (d, $^3J_{PC}$ = 6.1 Hz, Ar-C_q), 148.1 (Ar-C_q), 146.0 (d, J_{PC} = 5.7 Hz, Ar-C_q), 140.5 (d, J_{PC} = 3.1 Hz, Ar-*C*), 139.4 (Ar-*C*H), 133.6 (Ar-*C*H), 132.5 (d, J_{PC} = 9.2 Hz, Ar-CH), 132.2 (d, J_{PC} = 9.4 Hz, Ar-*C*H), 131.5 (d, J_{PC} = 2.7 Hz, Ar-*C*H), 131.1 (Ar-*C*H), 130.2 (Ar-*C*H), 129.0 (Ar-*C*H), 126.9 (Ar-*C*H), 126.7 (Ar-*C*H), 123.9 (d, J_{PC} = 3.5 Hz, Ar-*C*H), 123.5 (d, J_{PC} = 3.0 Hz, Ar-*C*H), 54.2 (Ph(*C*H)CH_3), 30.4 (d, $^3J_{PC}$ = 11.8 Hz, Ph(CH)*C*H_3), 28.9 (*C*H(CH_3)$_2$), 23.7 (CH(*C*H_3)$_2$).

^{31}P{^{1}H} NMR (C_6D_6, 162 MHz): δ [ppm] = 37.5

IR (ATR): $\tilde{\nu}$ [cm^{-1}] = 3058 (w), 2962 (m), 2948 (w), 1588 (m), 1457 (s), 1432 (s), 1382 (m), 1359 (m), 1335 (m), 1316 (m), 1285 (m), 1256 (m), 1237 (s), 1210 (m), 1188 (m), 1116 (s), 1098 (m), 1078 (s), 1057 (m), 1043 (m), 996 (m), 988 (m), 932 (m), 815 (m), 786 (vs), 755 (m), 744 (m), 725 (m), 716 (s), 698 (vs), 670 (m), 599 (m), 548 (s), 519 (m), 500 (m), 477 (m), 400 (m).

4.5.3 Synthesis of [{(*R*)-NEPIA}$_2$Cu$_2$] (17)

Following the similar procedure described above for **15**, the reaction of CuCl (71 mg, 0.71 mmol, 1.0 eq.), (*R*)-KNEPIA (400.0 mg, 0.74 mmol, 1.0 eq.) afforded single crystals from toluene suitable for X-ray analysis. The solvent was decanted, and the product was washed with cold *n*-pentane (5 mL).

Yield (based on crystals): 250 mg (60 %). **Elemental analysis** calcd. (%) for [$C_{72}H_{64}Cu_2N_4P_2$] (1174.34): C 73.64, H 5.49, N 4.77; found: C 74.07, H 5.85, N 4.86.

^{1}H NMR (C_6D_6, 300 MHz): δ [ppm] = 9.01 (d, $^3J_{HH}$ = 6.7 Hz, 4 H, *o*-Ar$_{phos}$-C*H*), 7.91-7.86 (m, 8 H, Ar$_{phos}$-C*H*), 7.68-7.66 (m, 4 H, Ar-C*H*), 7.61-7.59 (m, 4 H, Ar-C*H*), 7.53-7.48 (m, 8 H, Ar-C*H*), 7.27-7.23 (m, 4 H, Ar-C*H*), 7.14-7.09 (m, 5 H, Ar-C*H*), 6.81-6.72 (m, 11 H, Ar-C*H*), 5.12 (dq, $^3J_{HH}$ = 6.4 Hz, $^3J_{PH}$ = 16.3 Hz, 4 H, naph(C*H*)CH$_3$), 1.9 (d, $^3J_{PH}$ = 6.5 Hz, $^3J_{HH}$ = 1.7 Hz, 12 H, naph(CH)C*H*$_3$).

^{13}C{^{1}H} NMR (C_6D_6, 75 MHz): δ [ppm] = 145.6 (d, $^3J_{PC}$ = 3.6 Hz, Ar-C_q), 134.0 (Ar-*C*), 133.7 (Ar-*C*), 132.9 (d, $^3J_{PC}$ = 8.9 Hz, Ar-CH), 132.7 (Ar-CH), 131.2 (Ar-CH), 130.4 (Ar-CH), 128.9 (Ar-CH), 127.9 (Ar-CH), 127.7 (Ar-CH), 126.4 (d, J_{PC} = 20.5 Hz, Ar-CH), 125.1 (d, J_{PC} = 40.9 Hz Ar-*C*), 124.0 (*Ar*-CH), 123.3 (*Ar-C*), 50.6 (d, J_{PC} = 5.3 Hz, naph(*C*H)CH$_3$), 31.3 (d, $^3J_{PC}$ = 13.8 Hz, naph(CH)*C*H$_3$).

^{31}P{^{1}H} NMR (C_6D_6, 121 MHz): δ [ppm] = 40.2

IR (ATR): $\tilde{\nu}$ [cm^{-1}] = 3050 (m), 2963 (m), 2949 (vw), 1595 (m), 1510 (m), 1479 (vw), 1435 (m), 1393 (m), 1378 (m), 1363 (m), 1317 (m), 1302 (m), 1255 (m), 1226 (m), 1166 (m), 1141 (vs), 1117 (m), 1106 (s), 1100 (m), 1075 (s), 997 (s), 968 (m), 947 (m), 884 (m), 856 (vs), 839 (s), 796 (m), 776 (m), 746 (s), 715 (m), 694 (s), 653 (vs), 631 (s), 621 (s), 609 (vs), 579 (s), 554 (m), 521 (m), 508 (m), 496 (m), 480 (vs), 467 (s), 453 (m), 425 (m).

4.5.4 Synthesis of [{DippPIA}$_2$Cu$_2$] (18)

Following the similar procedure described above for **15** the reaction of CuCl (74.0 mg, 0.74 mmol, 1.0 eq.), HDippPIA (400.0 mg, 0.74 mmol, 1.0 eq.) and KN(SiMe$_3$)$_2$ (149.0 mg, 0.74 mmol, 1.0 eq.) afforded single crystals from toluene suitable for X-ray analysis. The solvent was decanted, and the product was washed with cold *n*-pentane (5 mL).

Yield (based on crystals): 248 mg (56 %). **Elemental analysis** calcd. (%) for [$C_{72}H_{88}Cu_2N_4P_2$] (1198.56): C 72.15, H 7.40, N 4.67; found: C 72.65, H 6.70, N 4.67.

^{1}H NMR (C_6D_6, 300 MHz): δ [ppm] = 9.94-9.89 (m, 2 H, Ar-C*H*), 7.73-7.68 (m, 2 H, Ar-C*H*), 7.36 (m, 2 H, Ar-C*H*), 7.13-7.02 (m, 12 H, Ar-C*H*), 6.96-6.92 (m, 4 H, Ar-C*H*), 6.80-6.77 (m, 2 H, Ar-C*H*), 6.66-6.62 (m, 6 H, Ar-C*H*), 6.34-6.29 (m, 2 H, Ar-C*H*), 4.45-4.37 (m, 4 H, C*H*(CH_3)$_2$), 3.66-3.59 (m, 4 H, *CH(CH$_3$)*$_2$), 1.82 (d, $^3J_{HH}$ = 6.75, 12 H, CH(C*H*$_3$)$_2$), 1.12 (d, $^3J_{HH}$ = 6.86, 12 H, CH(C*H*$_3$)$_2$), 0.47 (d, $^3J_{HH}$ = 6.82, 12 H, CH(C*H*$_3$)$_2$), 0.35 (d, $^3J_{HH}$ = 6.70, 12 H, CH(C*H*$_3$)$_2$).

^{13}C{^{1}H} NMR ($CDCl_3$, 75 MHz): δ [ppm] = 146.1 (Ar-C_q), 145.6 (d, J_{PC} = 5.7 Hz, Ar-C_q), 141.9 (Ar-*C*q), 133.2 (d, J_{PC} = 6.9 Hz, Ar-C_q), 127.1 (d, $^3J_{PC}$ = 10.6 Hz, Ar-*C*H), 124.4 (Ar-*C*H), 123.5 (d, $^3J_{PC}$ = 42.0 Hz, Ar-*C*H), 100.1 (*Ar-C*), 68.1 (*C*H(CH_3)$_2$), 25.8 (CH(*C*H$_3$)$_2$).

^{31}P{^{1}H} NMR (C_6D_6, 121 MHz): δ [ppm] = 8.4

IR (ATR): $\tilde{\nu}$ [cm^{-1}] = 3055 (w), 2959 (s), 2921 (s), 2861 (m), 1589 (w), 1479 (m), 1456 (m), 1428 (m), 1381 (m), 1358 (m), 1311 (m), 1258 (s), 1246 (vs), 1230 (s), 1190 (vs), 1158 (m), 1111 (m), 1099 (m), 1070 (m), 1044 (s), 998 (m), 964 (m), 932 (m), 918 (s), 883 (m), 827 (m), 791 (m), 744 (vs), 712 (m), 696 (m), 621 (m), 609 (s), 600 (m), 581 (m), 539 (m), 518 (s), 507 (s), 482 (m), 453 (m), 427 (m).

4.5.5 Synthesis of [{DippPIA}ZnPh] (19)

$ZnPh_2$ (147.0 mg, 0.28 mmol, 1.0 eq.) and HDippPIA (150.0 mg, 0.28 mmol, 1.0 eq.) was dissolved in toluene (30mL) and stirred for 16 h. The solvent was removed under reduced pressure and product was washed with *n*-pentane. The single crystals suitable for X-ray analysis was obtained from *n*-pentane. The solvent was decanted, and the product was washed with cold *n*-pentane (5 mL).

Yield (based on crystals): 120 mg (63 %). **Elemental analysis** calcd. (%) for [$C_{42}H_{49}ZnN_2P$] (678.22): C 74.38, H 7.28, N 4.13; found: C 73.68, H 6.78, N 4.16.

^{1}H NMR (C_6D_6, 400 MHz): δ [ppm] = 7.63-7.61 (m, 2 H, *o*-Ar$_{phos}$-C*H*), 7.33-7.28 (m, 4 H, Ar-C*H*), 7.14-7.12 (m, 9 H, Ar-CH), 6.92-6.77 (m, 6 H, Ar-C*H*), 3.88 (sept, $^3J_{HH}$ = 6.81 Hz, 4 H, C*H*(CH_3)$_2$), 1.03 (d, $^3J_{PH}$ = 6.87 Hz, 24 H, CH(C*H*$_3$)$_2$).

^{13}C{^{1}H} NMR (C_6D_6, 100 MHz): δ [ppm] = 146.9 (d, J_{PC} = 1.4 Hz, Ar-C_q), 146.3 (d, J_{PC} = 5.1 Hz, Ar-C_q), 140.9 (Ar-C_q), 139.5 (Ar-*C*H), 135.2 (Ar-*C*H), 134.3 (Ar-*C*H), 131.9 (d, J_{PC} = 8.7 Hz, Ar-*C*H), 131.2 (d, J_{PC} = 2.8 Hz, Ar-*C*H), 128.2 (Ar-*C*H), 127.8 (Ar-*C*H), 124.4 (d, J_{PC} = 2.8 Hz, Ar-*C*H), 124.0 (d, J_{PC} = 2.3 Hz, Ar-*C*H), 29.4 (*C*H(CH_3)$_2$), 24.1 (CH(*C*H$_3$)$_2$).

$^{31}P\{^1H\}$ NMR (C_6D_6, 162 MHz): δ [ppm] = 17.2.

IR (ATR): $\tilde{\nu}$ [cm^{-1}] = 3058 (w), 2962 (m), 2948 (m), 2922 (m), 2866 (m), 1588 (vw), 1457 (m), 1432 (s), 1382 (m), 1359 (m), 1335 (m), 1316 (s), 1285 (vs), 1256 (s), 1237 (s), 1210 (s), 1188 (s), 1116 (m), 1098 (m), 1078 (m), 1057 (m), 1043 (s), 996 (s), 988 (s), 932 (s), 815 (m), 786 (s), 755 (m), 744 (vs), 725 (m), 715 (m), 698 (vs), 670 (vs), 599 (m), 548 (m), 519 (m), 500 (s), 477 (s), 440 (m).

4.6 Synthesis and analytical data of group 13 metal complexes of chiral iminophosphonamides

4.6.1 Synthesis of [{(*R*)-PE^{Dipp}PIA}AlMe$_2$] (20)

To the pre-cold (at -40°C) solution of (*R*)-HPE^{Dipp}PIA (300 mg, 0.62 mmol, 1.0 eq) in 40 mL of toluene, $AlMe_3$ (0.25 mL, 0.62 mmol, 1.0 eq, 2M in heptane) was added dropwise. The resulting mixture was stirred overnight. After removal of all volatiles under reduced pressure and washing the residue with *n*-pentane (10 mL) afforded white solid powder. Single crystals suitable for X-ray structure analysis were obtained from hot *n*-heptane.

Yield (based on crystals): 220 mg (66%). **Elemental analysis** calcd. (%) for [$C_{34}H_{42}N_2PAl$] (536.68): C, 76.09, H, 7.89, N, 5.22; found: C, 76.36, H, 7.88, N, 5.19.

^1H NMR (C_6D_6, 400 MHz): δ [ppm] = 7.77-7.72 (m, 2 H, *o*-Ar-*H*), 7.29-7.27 (m, 2 H, *o*-Ar-*H*), 7.15-7.11 (m, 2 H, *o*-Ar-*H*), 7.08-6.99 (m, 9 H, Ar-*H*), 6.90-6.86 (m, 1 H, Ar-*H*), 6.78-6.74 (m, 2 H, Ar-*H*), 3.86 (dq, $^3J_{PH}$ = 17.1 Hz, $^3J_{HH}$ = 6.7 Hz, 1 H, Ph(C*H*)CH$_3$), 3.64 (br, 2 H, C*H*(CH$_3$)$_2$), 1.58 (dd, $^3J_{HH}$ = 6.71 Hz, $^4J_{PH}$ = 1.08 Hz, 3 H, Ph(CH)C*H*$_3$), 1.26-0.79 (br, 12 H , CH(C*H*$_3$)$_2$), 0.16 (s ,3 H, C*H*$_3$), 0.10 (s ,3 H, C*H*$_3$).

$^{13}C\{^1H\}$ NMR (C_6D_6, 100 MHz): δ [ppm] = 148.5 (d, J_{PC} = 4.7 Hz, Ar-*C*), 147.7 (d, J_{PC} = 5.4 Hz, Ar-*C*), 136.1 (d, J_{PC} = 4.4 Hz, Ar-*C*), 133.8 (d, J_{PC} = 9.8 Hz, Ar-*C*), 132.4 (d, J_{PC} = 9.8 Hz, Ar-*C*), 132.2 (d, J_{PC} = 2.8 Hz, Ar-*C*), 131.7 (d, J_{PC} = 2.9 Hz, Ar-*C*H), 131.6 (Ar-*C*H), 130.7 (Ar-*C*H), 129.3 (Ar-*C*H), 128.9 (Ar-*C*H), 128.7 (d, J_{PC} = 11.8 Hz, Ar-*C*H), 128.2 (Ar-*C*H), 127.9 (Ar-*C*H), 125.1 (d, J_{PC} = 3.6 Hz, Ar-*C*H), 124.2 (d, J_{PC} = 3.2 Hz, Ar-*C*H), 54.4 (Ph*C*H(CH$_3$)), 28.5 (*C*H(CH$_3$)$_2$), 27.7 (d, $^3J_{PC}$ = 12.9 Hz, PhCH(*C*H$_3$)), 22.7 (CH(*C*H$_3$)$_2$), 14.3 (CH(*C*H$_3$)$_2$), -4.4(Al-*C*H$_3$), -4.8(Al-*C*H$_3$).

$^{31}P\{^1H\}$ NMR (C_6D_6, 162 MHz): δ [ppm] = 36.3.

IR (ATR): $\tilde{\nu}$ [cm^{-1}] = 3060 (vw), 2963 (w), 2922 (w), 2865 (w), 1590 (w), 1487 (w), 1457 (m), 1435 (w), 1379 (w), 1364 (w), 1319 (w), 1273 (m), 1203 (m), 1170 (s), 1113 (s), 1104 (w), 1049 (m), 990 (s), 970 (m), 930 (w), 857 (s), 791 (m),752 (m), 718 (m), 694 (vs), 663 (s), 643 (m), 605 (m), 578 (m), 545 (m), 519 (s), 489 (m), 458 (w), 435 (vw).

4.6.2 Synthesis of [{(*R*)-PE^{Dipp}PIA}AlMe(C_6F_5)] (21)

To the pre-cold (at -40°C) solution of **20** (290.0 mg, 0.53 mmol, 1.0 eq) in 40 mL of toluene, $B(C_6F_5)_3$ (277.0 mg, 0.53 mmol, 1.0 eq) was added dropwise. The resulting mixture was stirred overnight. After removal of all volatiles under reduced pressure afforded colourless oil. Single crystals suitable for X-ray structure analysis were obtained from hot *n*-heptane.

Yield (based on crystals): 150 mg (22 %). **Elemental analysis** calcd. (%) for [$C_{39}H_{39}N_2PAlF_5$] (688.70): C, 68.02, H, 5.71, N, 4.07; found: C, 67.71, H, 5.08, N, 4.11.

^{1}H NMR (C_6D_6, 400 MHz): δ [ppm] = 7.61-7.56 (m, 2 H, *o*-Ar-*H*), 7.25-7.20 (m, 2 H, *o*-Ar-*H*), 7.13-7.11 (m, 2 H, *o*-Ar-*H*), 7.05-6.92 (m, 8 H, Ar-*H*), 6.88-6.79 (m, 4 H, Ar-*H*), 4.10-4.02 (m, 1 H, Ph(C*H*)CH$_3$), 3.37 (br, 2H, C*H*(CH$_3$)$_2$), 1.63 (dd, $^3J_{HH}$ = 6.72 Hz, $^4J_{PH}$ = 1.23 Hz, 3 H, Ph(CH)C*H*$_3$), 1.48-0.45 (m, 12 H, CH(C*H*$_3$)$_2$), 0.05 (s, 3 H, Al-C*H*$_3$).

^{13}C{^{1}H} NMR (C_6D_6, 100 MHz): δ [ppm] = 151.3 (d, J_{PC} = 22.5 Hz, Ar-C_q), 149.0 (d, J_{PC} = 29.4 Hz, Ar-C_q), 147.5 (d, J_{PC} = 5.4 Hz, Ar-C_q), 146.2 (d, J_{PC} = 8.0 Hz, Ar-C_q), 142.6 (Ar-*C*H), 138.6 (Ar-*C*H), 136.3 (Ar-*C*), 134.8 (d, J_{PC} = 4.6 Hz, Ar-*C*H), 133.8 (d, J_{PC} = 9.9 Hz, Ar-*C*H), 133.3 (d, J_{PC} = 10.0 Hz, Ar-*C*H), 132.8 (Ar-*C*H), 132.6 (Ar-*C*H), 132.1 (Ar-*C*H), 129.1 (Ar-*C*H), 128.9 (Ar-*C*H), 128.8 (Ar-*C*H), 128.6 (Ar-*C*H), 127.4 (Ar-*C*H), 125.6 (Ar-*C*H), 124.1 (Ar-*C*H), 54.6 (Ph*C*H(CH$_3$)), 28.5 (*C*H(CH$_3$)$_2$), 26.0 (d, $^3J_{PC}$ = 10.4 Hz, PhCH(*C*H$_3$)), 24.2 (CH(*C*H$_3$)$_2$), -4.9 (Al-*C*H$_3$).

^{19}F{^{1}H} NMR (C_6D_6, 376 MHz): δ [ppm] = -120.3 (m, *o*-C_6F_5), -155.3 (m, *p*-C_6F_5), -162.7 (m, *m*-C_6F_5).

^{31}P{^{1}H} NMR (C_6D_6, 162 MHz): δ [ppm] = 38.8.

IR (ATR): $\tilde{\nu}$ [cm^{-1}] = 3060 (vw), 2964 (w), 2863 (w), 1637 (m), 1590 (w), 1533 (w), 1506 (s), 1463 (vs), 1448 (m), 1434 (w), 1383 (m), 1370 (m), 1350 (m), 1322 (m), 1303 (s), 1281 (m), 1264 (m), 1207 (m), 1192 (s), 1162 (m), 1153 (m), 1119 (m), 1106 (m), 1072 (m), 1054 (m), 1028 (m), 1004 (m), 983 (s), 955 (s), 867 (s), 856 (m), 796 (m), 755 (m), 745 (m), 721 (s), 707 (s), 691 (m), 670 (s), 651 (m), 642 (m), 628 (m), 605 (s), 592 (m), 551 (m), 531 (s), 518 (m), 511 (s), 497 (s), 485 (m), 456 (m), 433 (m).

4.6.3 Synthesis of [{(*R*)-PE^{Dipp}PIA}$AlCl_2$] (22)

To the mixture of (*R*)-LiPE^{Dipp}PIA (300.0 mg, 0.62 mmol, 1.0 eq) and $AlCl_3$ (82.0 mg, 0.62 mmol, 1.0 eq) 40 mL of diethyl ether was added. The resulting solution was stirred overnight. After filtration and concentrating the filtrate afforded colourless crystals by hot heptane, suitable for X-ray structure analysis at room temperature. The solvent was decanted, and the product was washed with *n*-pentane (5 mL).

Yield (based on crystals): 170 mg (42 %). **Elemental analysis** calcd. (%) for [$C_{32}H_{36}N_2PAlCl_2$] (577.51): C 66.55, H 6.28, N 4.85; found: C 66.67, H 6.16, N 4.91.

^{1}H NMR (C_6D_6, 400 MHz): δ [ppm] = 7.86-7.80 (m, 2 H, Ar-*H*), 7.28-7.26 (m, 2 H, Ar-*H*), 7.07-6.85 (m, 14 H, Ar-*H*), 3.89-3.24 (m, 3 H, C*H*(CH_3)$_2$ & Ph(C*H*)CH_3), 1.79 (dd, $^3J_{HH}$ = 6.75 Hz, $^4J_{PH}$ = 1.57 Hz, 3 H, Ph(CH)C*H*$_3$), 1.61-0.03 (m, 12 H, CH(C*H*$_3$)$_2$).

^{13}C{^{1}H} NMR (C_6D_6, 100 MHz): δ [ppm] = 148.1 (d, J_{PC} = 4.9 Hz, Ar-*C*$_q$), 147.0 (d, J_{PC} = 2.8 Hz, Ar-*C*$_q$), 134.4 (d, J_{PC} = 10.4 Hz, Ar-*C*H), 133.3 (d, J_{PC} = 2.9 Hz, Ar-*C*H), 132.5 (d, J_{PC} = 10.4 Hz, Ar-*C*H), 129.13 (Ar-*C*H), 129.08 (Ar-*C*H), 129.0 (Ar-*C*H), 128.8 (Ar-*C*H), 128.5 (Ar-*C*H), 127.5 (Ar-*C*H), 127.2 (Ar-*C*H), 126.3 (Ar-*C*H), 126.0 (Ar-*C*H), 125.0 (Ar-*C*H), 124.5 (d, J_{PC} = 3.0 Hz, Ar-*C*H), 54.6 (Ph(*C*H)CH_3), 28.8 (*C*H(CH_3)$_2$), 27.4 (d, $^3J_{PC}$ = 13.5 Hz, Ph(CH)*C*H$_3$), 26.6 (CH(*C*H$_3$)$_2$).

^{31}P{^{1}H} NMR (C_6D_6, 162 MHz): δ [ppm] = 40.0.

IR (ATR): $\tilde{\nu}$ [cm^{-1}] = 3080 (vw), 2964 (m), 2926 (m), 2867 (m), 1588 (m), 1458 (s), 1436 (m), 1381 (m), 1361 (w), 1321 (w), 1255 (m), 1207 (m), 1185 (vs), 1155 (m), 1117 (s), 1106 (s), 1086 (s), 1045 (s), 1026 (s), 999 (s), 975 (s), 930 (m), 869 (m), 837 (s), 794 (s), 749 (m), 724 (vs), 695 (vs), 650 (m), 619 (m), 599 (m), 551 (m), 521 (s), 504 (m), 471 (m), 451 (m), 433 (w).

4.6.4 Synthesis of [{(*R*)-PEPIA}$_2$AlCl] (23)

To the mixture of (*R*)-LiPEPIA (1.0 g, 2.32 mmol, 2.0 eq.) and $AlCl_3$ (155 mg, 1.16 mmol, 1.0 eq.), 60 mL of diethyl ether was added. The resulting solution was stirred overnight. After filtration and concentrating the filtrate afforded colourless crystals, suitable for X-ray structure analysis at room temperature. The mother liquor was decanted, and the crystals were washed with *n*-pentane (5 mL).

Yield (based on crystals): 750 mg (71%). **Elemental analysis** calcd. (%) for [$C_{56}H_{56}N_4P_2AlCl$] (909.47): C 73.96, H 6.21, N 6.16; found: C 74.47, H 6.67, N 5.67.

^{1}H NMR (C_6D_6, 400 MHz, 298 K): δ [ppm] = 7.68 (br, 8 H, Ar-*H*), 7.09-6.84 (m, 32 H, Ar-*H*), 5.17 (br, 4 H Ph(C*H*)CH$_3$), 1.89 (br, 12 H, Ph(CH)C*H*$_3$). **^{1}H** NMR (thf-*d*$_8$, 400 MHz, 298 K): δ [ppm] = 7.51 (br, 8 H, Ar-*H*), 7.31 (br, 4 H, Ar-*H*), 7.13 (br, 8 H, Ar-*H*), 6.93-6.79 (br, 20 H, Ar-*H*), 4.79 (br, 4 H Ph(C*H*)CH$_3$), 1.59 (d, $^3J_{HH}$ = 6.58 Hz, 12 H, Ph(CH)C*H*$_3$).

^{13}C{^{1}H} NMR (C_6D_6, 100 MHz): δ [ppm] = 146.8 (Ar-*C*$_q$), 134.5-134.4 (Ar-*C*H), 131.1 (Ar-*C*H), 130.1 (Ar-*C*H), 128.2-127.9 (Ar-*C*H), 127.7 (Ar-*C*H), 127.5 (d, J_{PC} = 11.9 Hz, Ar-*C*H), 126.0 (Ar-*C*H), 54.3 (Ph(*CH*)CH$_3$), 26.4 (Ph(C*H*)*C*H$_3$).

^{31}P{^{1}H} NMR (C_6D_6, 162 MHz): δ [ppm] = 35.6.

IR (ATR): $\tilde{\nu}$ [cm^{-1}] = 3057 (vw), 2989 (vw), 2960 (w), 2860 (vw), 1496 (w), 1484 (vw), 1452 (w), 1439 (m), 1434 (m), 1372 (w), 1364 (w) ,1215 (w), 1205 (w), 1190 (w), 1185 (vs), 1128 (s), 1114 (s), 1094 (m), 1068 (m) ,1045 (m), 1039 (m), 1025 (m), 998 (w), 974 (vs), 907 (m), 858 (m), 837 (m), 823 (m), 776 (w), 759 (m), 752 (m), 743 (m), 734 (w), 730 (w), 722 (w), 706 (vs), 663 (m), 641 (m), 617 (w), 590 (m), 579 (s), 539 (m), 516 (vs), 487 (m), 470 (m), 456 (m), 414 (m).

4.6.5 Synthesis of [{(*R*)-PEPIA}$_2$GaCl] (24)

Following the procedure described above for **23** the reaction of (*R*)-LiPEPIA (1.0 g, 2.32 mmol, 2.0 eq) and $GaCl_3$ (205 mg, 1.16 mmol, 1.0 eq) in 60 mL of diethyl ether, afforded colourless crystals, suitable for X-ray structure analysis at room temperature. The solvent was decanted, and the product was washed with *n*-pentane (5 mL).

Yield (based on crystals): 650 mg (59%). **Elemental analysis** calcd. (%) for [$C_{56}H_{56}N_4P_2GaCl$] (952.21): C 70.64, H 5.93, N 5.88; found: C 70.74, H 6.57, N 5.58.

^{1}H NMR (C_6D_6, 400 MHz): δ [ppm] = 7.69-7.65 (m, 8 H, Ar-*H*), 7.11-7.10 (m, 8 H, Ar-*H*), 7.01-6.85 (m, 24 H *o,m,p*-Ar-*H*), 5.10 (br, 4 H, Ph(C*H*)CH$_3$), 1.90 (d, $^3J_{HH}$ = 6.52 Hz, 12 H, Ph(CH)C*H*$_3$).

^{13}C{^{1}H} NMR (C_6D_6, 100 MHz): δ [ppm] = 147.0 (Ar-*C*$_q$), 134.2 (d, J_{PC} = 10.3 Hz, Ar-*C*H), 131.1 (Ar-*C*H), 130.1 (Ar-*C*H), 128.1 (Ar-*C*H), 127.7 (Ar-*C*H), 127.6 (d, J_{PC} = 12.0 Hz, Ar-*C*H), 126.0 (*Ar-C*H), 54.5 (Ph(*C*H)CH$_3$), 26.4 (Ph(CH)*C*H$_3$).

$^{31}P\{^1H\}$ NMR (C_6D_6, 162 MHz): δ [ppm] = 36.5.

IR (ATR): $\tilde{\nu}$ [cm^{-1}] = 3058 (vw), 3025 (vw), 2960 (vw), 2925 (vw), 1493 (w), 1451 (w), 1436 (m), 1372 (w), 1276 (w), 1238 (m), 1205 (s), 1182 (m), 1142 (vw), 1129 (vw), 1113 (w), 1104 (w), 1068 (vw), 1043 (w), 999 (w), 868 (m), 829 (m), 772 (m), 749 (s), 743 (m), 693 (vs), 658 (w), 633 (w), 570 (m), 530 (vs), 508 (m), 446 (w).

4.6.6 Synthesis of [{(*R*)-PEPIA}$_2$Al]$^+$[$GaCl_4$]$^-$ (25)

To the mixture of **23** (150 mg, 0.33 mmol, 1.0 eq) and $GaCl_3$ (29 mg, 0.33 mmol, 1.0 eq), 10 mL of thf was added at room temperature. All the volatiles were removed *in vacuo* after stirring for 2 hours to obtain white powder. Single crystals suitable for X-ray structure analysis were grown from concentrated benzene solution at room temperature.

Yield (based on crystals): 98 mg (55%). **Elemental analysis** calcd. (%) for [$C_{56}H_{56}N_4P_2AlGaCl_4$] (1085.54): C 61.96, H 5.20, N 5.16; found: C 62.69, H 5.25, N 5.08.

^1H NMR ($CDCl_3$, 400 MHz): δ [ppm] = 7.64-7.60 (m, 4 H ,Ar-*H*), 7.42-7.38 (m, 8 H, Ar-*H*), 7.35-7.30 (m, 8 H, Ar-*H*), 7.17-7.13 (m, 4H, Ar-*H*), 7.10-7.06 (m, 8H, Ar-*H*), 7.01-6.99 (m, 8H, Ar-*H*), 4.35-4.27 (m, 4H, Ph(C*H*)CH$_3$), 1.56 (dd, $^3J_{HH}$ = 6.67 Hz, $^4J_{PH}$ = 1.44 Hz, 12 H, Ph(CH)C*H*$_3$).

$^{13}C\{^1H\}$ NMR ($CDCl_3$, 100 MHz): δ [ppm] = 143.9 (d, J_{PC} = 3.9 Hz, Ar-C_q), 134.0 (d, J_{PC} = 2.9 Hz, Ar-CH), 133.0 (d, J_{PC} = 11.2 Hz, Ar-CH), 129.2 (d, J_{PC} = 12.8 Hz, Ar-CH), 128.8 (Ar-CH), 127.8 (Ar-CH), 126.7 (Ar-CH), 124.7 (d, J_{PC} = 99.5 Hz, Ar-CH), 53.6 (Ph(CH)CH$_3$), 27.8 (d, $^3J_{PC}$ = 12.8 Hz, Ph(CH)CH$_3$).

$^{31}P\{^1H\}$ NMR (thf-d$_8$, 162 MHz): δ [ppm] = 43.3.

IR (ATR): $\tilde{\nu}$ [cm^{-1}] = 3027 (vw), 2976 (vw), 2923 (vw), 1589 (w), 1493 (w), 1455 (m), 1437 (w), 1371 (vw), 1313 (w), 1277(m), 1208(w), 1139 (s), 1124 (m), 1111 (w), 1069 (m), 1045 (w), 1000 (w), 974 (vw), 883 (w), 857 (vs), 766 (m), 752 (w), 747 (m), 693 (vs), 651 (vw),617 (w), 599 (w), 518 (vs), 478 (m), 439 (vw).

4.6.7 Synthesis of [{(*R*)-PEPIA}$_2$Ga]$^+$[$AlCl_4$]$^-$ (26)

Following the procedure described above for **25**, the reaction of **24** (150 mg, 0.16 mmol, 1.0 eq) and $AlCl_3$ (7 mg, 0.16 mmol, 1.0 eq) afforded white powder. Single crystals suitable for X-ray structure analysis were grown from concentrated benzene solution at room temperature.

Yield (based on crystals): 104 mg (60%). **Elemental analysis** calcd. (%) for [$C_{56}H_{56}N_4P_2AlGaCl_4$] (1085.54): C 61.96, H 5.20, N 5.16; found: C 61.98, H 5.17, N 5.22.

^{1}H NMR ($CDCl_3$, 400 MHz): δ [ppm] = 7.64-7.60 (m, 4 H ,Ar-*H*), 7.44-7.31 (m, 16 H, Ar-*H*), 7.18-7.10 (m, 12 H, Ar-*H*), 7.02-7.00 (m, 8H, Ar-*H*), 4.31 (dq, $^3J_{PH}$ = 9.42 Hz, $^3J_{HH}$ = 6.63 Hz 4H, Ph(C*H*)CH$_3$), 1.55 (dd, $^3J_{HH}$ = 6.71 Hz, $^4J_{PH}$ = 1.85 Hz, 12 H, Ph(CH)C*H*$_3$).

^{13}C{^{1}H} NMR ($CDCl_3$, 100 MHz): δ [ppm] = 143.9 (d, J_{PC} = 3.8 Hz, Ar-C_q), 133.9 (Ar-*C*H), 132.9 (d, J_{PC} = 11.3 Hz, Ar-*C*H), 129.2 (d, J_{PC} = 12.8 Hz, Ar-*C*H), 128.8 (Ar-*C*H), 127.9 (Ar-*C*H), 126.8 (Ar-*C*H), 125.1 (d, J_{PC} = 99.6 Hz, Ar-*C*H) 54.4 (Ph(*C*H)CH$_3$), 27.9 (d, $^3J_{PC}$ = 12.4 Hz, Ph(CH)*C*H$_3$).

^{31}P{^{1}H} NMR ($CDCl_3$, 162 MHz): δ [ppm] = 48.6.

IR (ATR): $\tilde{\nu}$ [cm^{-1}] = 3062 (vw), 3029 (vw), 2970 (w), 2931 (w), 2845 (vw), 1589 (vw), 1493 (w), 1452 (w), 1436 (m), 1367 (w), 1278 (w), 1242 (w), 1205 (s), 1191 (s), 1140 (w), 1110 (w), 1102 (w), 910 (w), 828 (s), 775 (m), 758 (s), 746 (m), 721(m), 695 (vs), 664 (w), 633 (m), 616 (m), 568 (m), 558 (s), 518 (s), 506 (m), 449 (w).

4.6.8 Synthesis of [{(*R*)-PEPIA}$_2$AlH] (27)

$LiAlH_4$ used in the following synthesis was purified by extracting with diethyl ether followed by filtration and removal of the solvent from the filtrate under vacuum to obtain $LiAlH_4$ as white powder.

To the mixture of (*R*)-HPEPIA (425 mg, 1.0 mmol, 2.0 eq) and $LiAlH_4$ (19 mg, 0.5 mmol, 1.0 eq) 40 mL of diethyl ether was added at room temperature. The reaction mixture was stirred overnight. After filtration and storing the concentrated filtrate at -30 °C for 2 days afforded colourless crystals suitable for X-ray analysis. The mother liquor was decanted-off and the crystals were washed with *n*-pentane (5 mL) and dried under vacuum.

Yield (based on crystals): 285 mg (65%). **Elemental analysis** calcd. (%) for [$C_{56}H_{57}N_4P_2Al$] (875.03): C 76.87, H 6.57, N 6.40; found: C 76.93, H 6.97, N 6.20.

^{1}H NMR (C_6D_6, 400 MHz): δ [ppm] = 7.63 (br, 8 H, *o*-Ar$_{phos}$-*H*), 7.20-7.18 (m, 8 H, Ar-*H*), 7.00-6.85 (m, 24 H *o, m, p*-Ar-*H*), 4.99 (br, 4 H Ph(C*H*)CH$_3$), 1.90 (d, $^3J_{HH}$ = 6.45 Hz, 12 H, Ph(CH)C*H*$_3$).

$^{13}C\{^1H\}$ NMR (C_6D_6, 100 MHz): δ [ppm] = 147.8 (Ar-C_q), 133.9 (d, $^2J_{PC}$ = 10.2 Hz, *o*-Ar$_{phos}$-*C*H), 131.7 (d, J_{PC} = 91.0 Hz, Ar-*C*H), 130.8 (d, J_{PC} = 2.7 Hz, Ar-*C*H), 128.0 (Ar-*C*H), 127.7 (Ar-*C*H), 127.6 (Ar-*C*H), 125.9 (*Ar*-*C*H), 54.1 (Ph(*C*H)CH$_3$), 25.6 (d, $^3J_{PC}$ = 12.3 Hz, Ph(CH)*C*H$_3$)

$^{31}P\{^1H\}$ NMR (C_6D_6, 162 MHz): δ [ppm] = 33.6.

IR (ATR): $\tilde{\nu}$ [cm^{-1}] = 3081 (vw), 3054 (vw), 3024 (vw), 2969 (vw), 2958 (vw), 2929 (vw), 1693 (m), 1601 (w), 1590 (vw), 1493 (w), 1435 (w), 1369 (w), 1349 (w), 1273 (w), 1238 (w), 1220 (vs), 1202 (m), 1189 (w), 1136 (w), 1111 (m), 1103 (m), 1068 (w), 1044 (vw), 1021(vw), 984 (w), 973 (w), 905 (vw), 856 (s), 826 (s),783 (m), 771(m), 753 (m), 740 (m), 719 (vs), 689 (w) ,663 (m), 606 (m), 584 (w), 571 (s), 537 (s), 508 (w), 465 (vw), 418 (vw).

4.7 Synthesis of group 4 metal complexes of chiral iminophosphonamide:

4.7.1 Synthesis of [{(*R*)-PEPIA}Hf(NMe$_2$)$_3$] (**28**)

[Hf(NMe$_2$)$_4$] (76 mg, 0.21 mmol, 1.0 eq.) and (*R*)-HPEPIA (91 mg, 0.21 mmol, 1.0 eq.) was dissolved in toluene (20 mL) and stirred for 16 h at room temperature. The solvent was removed which afforded white powder. Colourless crystals suitable for X-ray analysis were obtained by storing the *n*-pentane solution at room temperature for overnight.

Yield (based on crystals): 92 mg (60 %). **Elemental analysis** calcd. (%) for [$C_{34}H_{46}HfN_5P$] (734.30): C 55.62, H 6.32, N 9.54; found: C 55.92, H 6.41, N 8.71.

^1H NMR (C_6D_6, 400 MHz): δ [ppm] = 7.55-7.49 (m, 4 H, Ar-C*H*), 7.14-7.11 (m, 4 H, Ar-C*H*), 7.04-6.91 (m, 12 H, Ar-C*H*), 4.43 (dq, $^3J_{HH}$ = 6.73 Hz, $^3J_{PH}$ = 17.62 Hz, 2 H, Ph(C*H*)CH$_3$), 3.3 (s, 18 H, N(Me$_2$)$_3$) 1.46 (d, 6 H, $^3J_{HH}$ = 7.39 Hz, Ph(CH)C*H*$_3$).

$^{13}C\{^1H\}$ NMR (C_6D_6, 75 MHz): δ [ppm] = 147.8 (d, J_{PC} = 5.6 Hz, Ar-C_q), 133.0 (d, J_{PC} = 9.6 Hz, Ar-C_q), 132.1 (Ar-*C*H), 131.4 (d, J_{PC} = 2.8 Hz, Ar-*C*H), 131.2 (Ar-*C*H), 128.0 (Ar-*C*H), 127.1 (Ar-*C*H), 126.2 (Ar-*C*H), 56.2 (Ph(*C*H)CH$_3$), 44.2 ((NMe$_2$)$_3$), 27.1 (d, J_{PC} = 12.3 Hz, Ph(CH)*C*H$_3$).

$^{31}P\{^1H\}$ NMR (C_6D_6, 121 MHz): δ [ppm] = 36.5

IR (ATR): $\tilde{\nu}$ [cm^{-1}] = 3058 (m), 2963 (m), 2810 (m), 2749 (w), 1600 (m), 1491 (m), 1451 (m), 1435 (m), 1367 (m), 1308 (m), 1277 (m), 1244 (vs), 1233 (m), 1206 (m), 1146 (m), 1135 (m), 1106 (s), 1084 (m), 1070 (m), 1027 (m), 999 (vs), 978 (s), 964 (m), 944 (m), 903 (m), 855 (m),

811 (vs), 779 (m), 757 (vs), 744 (m), 718 (vs), 694 (m), 653 (m), 623 (m), 608 (m), 574 (m), 544 (m), 517 (vs), 438 (m), 414 (m).

4.7.2 Synthesis of [{(*R*)-PE^{Dipp}PIA}Zr(NMe$_2$)$_3$] (29)

[Zr(NMe$_2$)$_4$] (111 mg, 0.42 mmol, 1.0 eq.) and (*R*)-HPE^{Dipp}PIA (200.0 mg, 0.42 mmol, 1.0 eq.) was dissolved in toluene (20 mL) and stirred for 16 h at room temperature. The solvent was removed which afforded a white powder. Colourless crystals suitable for X-ray analysis were obtained by storing the *n*-pentane solution at -10 °C for 2 days.

Yield (based on crystals): 207 mg (70 %). **Elemental analysis** calcd. (%) for [$C_{38}H_{54}ZrN_5P$] (703.08): C 64.92, H 7.74, N 9.96; found: C 65.67, H 7.37, N 9.42.

^{1}H NMR (C_6D_6, 400 MHz): δ[ppm] = 7.71-7.66 (m, 2 H, Ar-C*H*), 7.27-7.25 (m, 2 H, Ar-C*H*), 7.15-7.01 (m, 10 H, Ar-C*H*), 6.94-6.91 (m, 2 H, Ar-C*H*), 6.83-6.79 (m, 2 H, Ar-C*H*), 4.07 (dq, $^3J_{HH}$ = 6.72 Hz, $^3J_{PH}$ = 17.32 Hz, 1 H, Ph(C*H*)CH$_3$), 3.75-3.66 (m, 1 H, C*H*(CH$_3$)$_2$), 3.19-3.15 (br, 19 H, N(Me$_2$)$_3$ & C*H*(CH$_3$)$_2$), 1.47 (dd, 3 H, $^3J_{HH}$ = 6.72 Hz, $^4J_{PH}$ = 0.71, Ph(CH)C*H*$_3$), 1.33 (d, 3 H, $^3J_{HH}$ = 6.81 Hz, CH(C*H*$_3$)$_2$), 1.24 (d, 3 H, $^3J_{HH}$ = 6.83 Hz, CH(C*H*$_3$)$_2$), 0.48 (d, 3 H, $^3J_{HH}$ = 6.74 Hz, CH(C*H*$_3$)$_2$), 0.20 (d, 3 H, $^3J_{HH}$ = 6.69 Hz, CH(C*H*$_3$)$_2$).

^{13}C{^{1}H} NMR (C_6D_6, 75 MHz): δ[ppm] = 148.5 (d, J_{PC} = 6.5 Hz, Ar-C_q), 146.1 (d, J_{PC} = 5.8 Hz, Ar-C_q), 145.9 (d, J_{PC} = 5.8 Hz, Ar-C_q), 141.4 (d, J_{PC} = 4.3 Hz, Ar-C_q), 134.1 (Ar-CH), 133.5 (d, J_{PC} = 9.0 Hz, Ar-*C*H), 133.0 (d, J_{PC} = 9.1 Hz, Ar-*C*H), 131.4 (d, J_{PC} = 2.7 Hz, Ar-*C*H), 131.0 (d, J_{PC} = 2.8 Hz, Ar-*C*H), 130.2 (Ar-*C*H), 128.5 (*Ar*-CH), 127.7 (*Ar*-CH), 126.8 (*Ar*-CH), 124.4 (d, J_{PC} = 3.7 Hz, Ar-CH), 124.1 (d, J_{PC} = 3.2 Hz, Ar-*C*H), 123.9 (d, J_{PC} = 3.3 Hz, Ar-*C*H), 57.0 (Ph(*C*H)CH$_3$), 44.5 ((NMe$_2$)$_3$), 29.2 (*C*H(CH$_3$)$_2$), 28.8 (*C*H(CH$_3$)$_2$), 26.3 (d, J_{PC} = 13.2 Hz, Ph(CH)*C*H$_3$), 25.0 (CH(*C*H$_3$)$_2$), 24.7 (CH(*C*H$_3$)$_2$), 24.6 (CH(*C*H$_3$)$_2$), 23.8 (CH(*C*H$_3$)$_2$).

^{31}P{^{1}H} NMR (C_6D_6, 121 MHz): δ[ppm] = 33.9

IR (ATR): $\tilde{\nu}$ [cm^{-1}] = 3053 (m), 2963 (s), 2925 (m), 2864 (m), 2812 (m), 2760 (m), 1588 (w), 1493 (s), 1482 (vs), 1452 (s), 1433 (m), 1380 (m), 1359 (m), 1319 (m), 1297 (m), 1274 (m), 1257 (m), 1242 (m), 1234 (s), 1202 (s), 1157 (vs), 1126 (s), 1101 (m), 1054 (m), 1028 (m), 990 (vs), 959 (s), 939 (s), 840 (m), 790 (m), 779 (m), 746 (m), 716 (s), 695 (vs), 671 (m), 565 (m), 550 (m), 536 (s), 513 (m).

4.7.3 Synthesis of [{(*R*)-PE^{Dipp}PIA}Hf(NMe$_2$)$_3$] (**30**)

Following the similar procedure described above for **29**, the reaction of [Hf(NMe$_2$)$_4$] (149.0 mg, 0.42 mmol, 1.0 eq.) and (*R*)-HPE^{Dipp}PIA (200.0 mg, 0.42 mmol, 1.0 eq.) afforded crystals suitable for X-ray analysis by storing the pentane solution at -10 °C for 2 days.

Yield (based on crystals): 199 mg (60 %). **Elemental analysis** calcd. (%) for [$C_{38}H_{54}HfN_5P$] (790.4): C 57.75, H 6.89, N 8.86; found: C 57.83, H 7.42, N 8.14.

^{1}H NMR (C_6D_6, 400 MHz): δ[ppm] = 7.70-7.65 (m, 2 H, Ar-C*H*), 7.24-7.22 (m, 2 H, Ar-C*H*), 7.14-6.76 (m, 14 H, Ar-C*H*), 4.17 (dq, $^3J_{HH}$ = 6.73 Hz, $^3J_{PH}$ = 16.03 Hz, 1 H, Ph(C*H*)CH$_3$), 3.76-3.66 (m, 1 H, C*H*(CH$_3$)$_2$), 3.21 (br, 19 H, (NMe$_2$)$_3$ & C*H*(CH$_3$)$_2$), 1.52 (dd, 3 H, $^3J_{HH}$ = 6.75 Hz, $^4J_{PH}$ = 1.10, Ph(CH)C*H*$_3$), 1.33 (d, 3 H, $^3J_{HH}$ = 6.80 Hz, CH(C*H*$_3$)$_2$), 1.26 (d, 3 H, $^3J_{HH}$ = 6.83 Hz, CH(C*H*$_3$)$_2$), 0.47 (d, 3 H, $^3J_{HH}$ = 6.74 Hz, CH(C*H*$_3$)$_2$), 0.18 (d, 3 H, $^3J_{HH}$ = 6.72 Hz, CH(C*H*$_3$)$_2$).

^{13}C{^{1}H} NMR (C_6D_6, 75 MHz): δ[ppm] = 148.3 (d, $^3J_{PC}$ = 5.7 Hz, Ar-*C*$_q$), 146.4 (d, $^3J_{PC}$ = 5.7 Hz, Ar-*C*$_q$), 146.2 (d, $^3J_{PC}$ = 5.7 Hz, Ar-*C*$_q$), 141.0 (d, $^3J_{PC}$ = 4.2 Hz, Ar-*C*$_q$), 133.8 (Ar-*C*H), 133.6 (d, J_{PC} = 9.1 Hz, Ar-*C*H), 133.0 (d, J_{PC} = 9.2 Hz, Ar-*C*H), 131.6 (d, J_{PC} = 2.9 Hz, Ar-*C*H), 131.0 (d, J_{PC} = 2.8 Hz, Ar-*C*H), 130.2 (Ar-*C*H), 128.5 (Ar-*C*H), 127.7 (Ar-*C*H), 126.9 (*Ar*-*C*H), 124.7 (d, J_{PC} = 3.7 Hz, Ar-*C*H), 124.0 (d, J_{PC} = 3.2 Hz, Ar-*C*H), 123.9 (d, J_{PC} = 3.2 Hz, Ar-*C*H), 56.9 (Ph(*C*H)CH$_3$), 44.2 ((NMe$_2$)$_3$), 29.2 (*C*H(CH$_3$)$_2$), 28.8 (*C*H(CH$_3$)$_2$), 25.9 (d, J_{PC} = 3.2 Hz, Ph(CH)*C*H$_3$), 25.1 (CH(*C*H$_3$)$_2$), 24.8 (CH(*C*H$_3$)$_2$), 24.6 (CH(*C*H$_3$)$_2$), 23.9 (CH(*C*H$_3$)$_2$).

^{31}P{^{1}H} NMR (C_6D_6, 121 MHz): δ[ppm] = 33.7

IR (ATR): $\tilde{\nu}$ [cm^{-1}] = 3053 (w), 2970 (m), 2926 (m), 2865 (m), 2853 (m), 2813 (m), 2760 (m), 1589 (vw), 1493 (vw), 1462 (vw), 1452 (vw), 1434 (m), 1381 (m), 1361 (m), 1319 (m), 1308 (m), 1275 (m), 1247 (m), 1237 (s), 1198 (s), 1150 (m), 1127 (m), 1102 (s), 1058 (s), 1046 (m), 1028 (s), 991 (vs), 965 (s), 945 (s), 856 (m), 844 (m), 791 (m), 780 (m), 762 (m), 746 (m), 717 (m), 695 (m), 673 (m), 621 (m), 611 (m), 597 (m), 567 (m), 551 (m), 524 (s), 514 (vs), 480 (m), 452 (vs), 433 (m).

4.8 Synthesis of chiral homoleptic lanthanide complexes [{(*R*)-PEPIA}$_3$Ln] (Ln = Y (**31**), La (**32**), Tb (**33**), Yb (**34**) and Lu (**35**))

To the mixture of LnCl$_3$ (0.35 mmol, 1.4 eq.) and (*R*)-KPEPIA (347 mg, 0.75 mmol, 3.0 eq.) 30 mL of toluene was added. The resulting reaction mixture was refluxed for 16 h. The colourless precipitate was filtered off and the solvent was concentrated in vacuo. Single crystals suitable for X-ray crystallography were afforded from hot toluene. The crystals were washed with *n*-pentane.

4.8.1 Synthesis of [{(*R*)-PEPIA}$_3$Y] (**31**)

Yield (based on crystals): 184 mg (54 %). **Elemental analysis** calcd. (%) for [$C_{91}H_{92}YN_6P_3$] = {**31**·toluene} (1451.57): C 75.30, H 6.39, N 5.79; found: C 75.78, H 6.22, N 5.35.

^{1}H NMR (C_6D_6, 400 MHz): δ [ppm] = 7.53-7.50 (m, 12 H, Ar-C*H*), 7.22 (m, 18 H, Ar-C*H*), 6.89-6.86 (m, 6 H, Ar-C*H*), 6.74-6.67 (m, 24 H, Ar-H), 4.98 (q, $^3J_{HH}$ = 6.50 Hz, 6 H, Ph(C*H*)CH$_3$), 2.15 (dd, 18 H, $^3J_{HH}$ = 2.91 Hz, $^3J_{PH}$ = 6.48, Ph(CH)C*H*$_3$).

^{13}C{^{1}H} NMR (C_6D_6, 75 MHz): δ [ppm] = 147.5 (d, $^3J_{PC}$ = 5.1 Hz, Ar-C_q), 133.3 (d, J_{PC} = 86.3 Hz, Ar-C_q), 132.6 (d, J_{PC} = 8.7 Hz, Ar-*C*H), 129.4 (d, J_{PC} = 2.6 Hz, Ar-*C*H), 129.0 (Ar-*C*H), 127.4 (d, J_{PC} = 10.8 Hz, Ar-*C*H), 126.2 (Ar-*C*H), 125.7 (Ar-*C*H), 58.2 (Ph(*C*H)CH$_3$), 30.8 (d, J_{PC} = 21.6 Hz, Ph(CH)*C*H$_3$).

^{31}P{^{1}H} NMR (C_6D_6, 121 MHz): δ [ppm] = 30.3

IR (ATR): $\tilde{\nu}$ [cm^{-1}] = 3080 (vw), 3052 (vw), 3024 (vw), 1600 (w), 1492 (m), 1453 (m), 1436 (m), 1372 (m), 1360 (w), 1309 (m), 1274 (m), 1204 (m), 1153 (s), 1125 (m), 1099 (m), 1070 (m), 1048 (m), 1028 (m), 1003 (m), 983 (m), 909 (m), 833 (vs), 778 (m), 743 (s), 730 (s), 715 (vs), 695 (m), 633 (m), 616 (vw), 580 (m), 561 (m), 523 (s), 518 (vs), 465 (m), 436 (m), 433 (m).

4.8.2 Synthesis of [{(*R*)-PEPIA}$_3$La] (**32**)

Yield (based on crystals): 141 mg (40 %). **Elemental analysis** calcd. (%) for [$C_{84}H_{84}LaN_6P_3$] (1409.43): C 71.58, H 6.01, N 5.96; found: C 71.94, H 5.99, N 5.75.

^{1}H NMR (C_6D_6, 400 MHz): δ [ppm] = 7.62-7.58 (m, 12 H, Ar-C*H*), 7.14-7.02 (m, 30 H, Ar-C*H*), 7.89-6.79 (m, 18 H, Ar-C*H*), 4.85 (dq, $^3J_{HH}$ = 6.55 Hz, $^3J_{PH}$ = 13.21 Hz, 6 H, Ph(C*H*)CH$_3$), 2.11 (dd, 18 H, $^3J_{HH}$ = 6.62 Hz, $^4J_{PH}$ = 2.44, Ph(CH)C*H*$_3$).

^{13}C{^{1}H} NMR (C_6D_6, 75 MHz): δ [ppm] = 148.3 (d, $^3J_{PC}$ = 3.8 Hz, Ar-C_q), 134.1 (Ar-C_q), 133.6 (Ar-C_q), 133.1 (d, J_{PC} = 8.8 Hz, Ar-*C*H), 129.7 (d, J_{PC} = 2.6 Hz, Ar-*C*H), 127.5 (d, J_{PC} = 10.7 Hz, Ar-*C*H), 127.9 (Ar-*C*H), 126.0 (Ar-*C*H), 57.5 (Ph(*C*H)CH$_3$), 31.0 (d, J_{PC} = 20.7 Hz, Ph(CH)*C*H$_3$).

^{31}P{^{1}H} NMR (C_6D_6, 121 MHz): δ [ppm] = 30.1

IR (ATR): $\tilde{\nu}$ [cm^{-1}] = 3054 (m), 3023 (s), 2960 (m), 2915 (m), 2845 (vw), 1673 (m), 1594 (m), 1486 (s), 1436 (m), 1394 (m), 1356 (w), 1310 (s), 1273 (s), 1234 (s), 1174 (s), 1141 (m), 1100 (s), 1064 (m), 1026 (m), 979 (m), 934 (m), 855 (m), 809 (w), 775 (w), 744 (m), 695 (m), 607 (vs), 544 (vs), 520 (s).

4.8.3 Synthesis of [{(*R*)-PEPIA}$_3$Tb] (33)

Yield (based on crystals): 250 mg (30 %). **Elemental analysis** calcd. (%) for [$C_{84}H_{84}TbN_6P_3$] (1429.45): C 70.58, H 5.92, N 5.88; found: C 70.66, H 5.68, N 5.03.

IR (ATR): $\tilde{\nu}$ [cm^{-1}] = 3056 (vw), 3024 (w), 2969 (vw), 2846 (vw), 1601 (w), 1492 (m), 1453 (m), 1436 (m), 1371 (m), 1359 (m), 1309 (vw), 1274 (m), 1204 (w), 1156 (vs), 1126 (vs), 1106 (m), 1099 (m), 1069 (m), 1048 (m), 1028 (m), 1003 (m), 982 (vs), 946 (m), 909 (s), 831 (vs), 776 (s), 744 (s), 729 (vs), 715 (m), 694 (vs), 633 (m), 616 (m), 580 (m), 560 (m), 524 (m), 517 (m), 485 (m), 436 (m), 424 (m).

4.8.4 Synthesis of [{(*R*)-PEPIA}$_3$Yb] (34)

Yield (based on crystals): 126 mg (35 %). **Elemental analysis** calcd. (%) for [$C_{91}H_{92}YbN_6P_3$] = **34**·2 toluene] (1535.7): C 71.17, H 6.04, N 5.47; found: C 71.89, H 6.06, N 5.11.

$^{31}P\{^1H\}$ NMR (C_6D_6, 121 MHz): δ [ppm] = -89.6

IR (ATR): $\tilde{\nu}$ [cm^{-1}] = 3058 (w), 3024 (vw), 2967 (w), 2856 (w), 1600 (w), 1491 (m), 1452 (m), 1436 (m), 1372 (m), 1310 (m), 1274 (m), 1202 (s), 1154 (vs), 1122 (m), 1104 (m), 1098 (m), 1070 (m), 1048 (m), 1028 (m), 1003 (w), 985 (vw), 909 (vs), 835 (m), 776 (s), 742 (s), 715 (vs), 694 (vs), 634 (m), 616 (m), 580 (m), 561 (m), 524 (vs), 517 (vs), 465 (vw), 436 (vw), 433 (vw).

4.8.5 Synthesis of [{(*R*)-PEPIA}$_3$Lu] (35)

Yield (based on crystals): 163 mg (45 %). **Elemental analysis** calcd. (%) for [$C_{98}H_{100}LuN_6P_3$] = {**37**·2 toluene} (1174.34): C 72.22, H 6.18, N 5.16; found: C 72.14, H 6.14, N 5.16.

^1H NMR (C_6D_6, 400 MHz): δ [ppm] = 7.54-7.51 (m, 12 H, Ar-C*H*), 7.23-7.22 (m, 18 H, Ar-C*H*), 7.07-7.00 (m, 5 H, Ar-C*H*), 6.89-6.86 (m, 5 H, Ar-C*H*), 6.74-6.69 (m, 20 H, Ar-C*H*), 5.02 (q, $^3J_{HH}$ = 6.56 Hz, 6 H, Ph(C*H*)CH$_3$), 2.14 (dd, $^3J_{PH}$ = 6.76 Hz, $^3J_{HH}$ = 3.06 Hz, 18 H, Ph(CH)C*H*$_3$).

$^{13}C\{^1H\}$ NMR (C_6D_6, 75 MHz): δ [ppm] = 147.8 (d, $^3J_{PC}$ = 4.9 Hz, Ar-C_q), 133.3 (d, $^3J_{PC}$ = 9.6 Hz, Ar-C_q), 132.4 (d, J_{PC} = 7.6 Hz, Ar-CH), 131.9 (Ar-CH), 131.1 (d, J_{PC} = 2.6 Hz, Ar-CH), 128.2 (Ar-

*C*H), 127.2 (Ar-*C*H), 126.4 (d, J_{PC} = 42.2 Hz, Ar-*C*H), 55.8 (Ph(*C*H)CH$_3$), 28.9 (d, J_{PC} = 14.1 Hz, Ph(CH)*C*H$_3$).

^{31}P{^{1}H} NMR (C_6D_6, 121 MHz): δ [ppm] = 30.3

IR (ATR): $\tilde{\nu}$ [cm^{-1}] = 3051 (w), 3023 (vw), 2973 (w), 2955 (vw), 2913 (vw), 2853 (vw), 1597 (w), 1489 (m), 1450 (m), 1434 (m), 1358 (m), 1344 (m), 1314 (m), 1303 (m), 1274 (m), 1236 (s), 1182 (m), 1171 (s), 1148 (m), 1137 (m), 1101 (m), 1064 (m), 1027 (m), 1003 (m), 978 (m), 934 (m), 907 (s), 859 (m), 831 (m), 807 (m), 760 (m), 756 (s), 743 (m), 696 (vs), 632 (m), 608 (m), 543 (vs), 523 (s), 497 (m), 478 (m), 436 (m).

4.9 Synthesis and analytical data of chiral group 14 compounds

4.9.1 Synthesis of [{PhC(N^{t}Bu)$_2$}Si{N(*R*)(CH)MePh(PPh$_2$)}{=NDipp}] (36)

[{(*R*)PE^{Dipp}PIA}Li]$_2$ (441.0 mg, 0.45 mmol, 0.5 eq.) and [{PhC(N^{t}Bu)$_2$}SiCl] (267.0 mg, 0.90 mmol, 1.00 eq.) were dissolved in toluene (30 mL) in a one pot and stirred at room temperature for 16 h. After filtration and removal of the solvent under reduced pressure gave crystalline solid. The resulting white solid was washed with *n*-pentane (5 mL). Single crystals suitable for X-ray analysis were obtained from *n*-heptane. The solvent was decanted, and the product was washed with cold *n*-pentane (5 mL).

Yield (based on crystals): 250 mg (37.5 %). **Elemental analysis** calcd. (%) for [$C_{47}H_{60}N_4PSi$] (740.08): C 76.28, H 8.17, N 7.57; found: C 75.61, H 7.76, N 7.46.

^{1}H NMR (C_6D_6, 400 MHz): δ [ppm] = 7.73-7.69 (m, 2 H, Ar-C*H*), 7.65-7.63 (m, 2 H, Ar-C*H*), 7.54-7.53 (m, 1 H, Ar-C*H*), 7.38-7.36 (m, 2 H, Ar-C*H*), 7.30-7.25 (m, 3 H, Ar-C*H*), 7.13-6.85 (m, 13 H, Ar-C*H*), 5.63 (q, $^3J_{HH}$ = 7.28 Hz, 1 H, Ph(C*H*)CH$_3$), 4.18 (sept, $^3J_{HH}$ = 6.8 Hz, 2 H, C*H*(CH$_3$)$_2$), 2.14 (d, $^3J_{HH}$ = 7.28 Hz, 3 H, Ph(CH)C*H*$_3$), 1.43 (d, $^3J_{HH}$ = 6.85 Hz, 6 H, CH(C*H*$_3$)$_2$), 1.37 (d, $^3J_{HH}$ = 6.85 Hz, 6 H, CH(C*H*$_3$)$_2$), 1.18 (s, 18H, C(C*H*$_3$)$_3$).

^{13}C{^{1}H} NMR (C_6D_6, 100 MHz): δ [ppm] = 177.8 (d, J_{PC} = 2.7 Hz, ^{tBu}NCN), 146.5 (Ar$_{phos}$-*C*$_q$), 145.8 (Ar-*C*$_q$), 140.7 (d, J_{PC} = 19.7 Hz, Ar-*C*$_q$), 140.2 (Ar-*C*H), 140.1 (Ar-*C*H), 134.5 (d, J_{PC} = 22.0 Hz, *Ar*-*C*H), 133.3 (d, J_{PC} = 21.7 Hz, Ar-*C*H), 131.4 (*Ar*-*C*H), 130.5 (*Ar*-*C*H), 129.2 (*Ar*-*C*H), 129.1 (*Ar*-*C*H), 128.7 (*Ar*-*C*H), 127.5 (Ar-*C*H), 126.1 (Ar-*C*H), 123.2 (Ar-*C*H), 116.1 (Ar-*C*H), 59.7 (d, J_{PC} = 5.9 Hz, Ph(*C*H)CH$_3$), 54.8 (*C*(CH$_3$)$_3$), 54.7 (*C*(CH$_3$)$_3$), 31.6 (*C*H(CH$_3$)$_2$), 31.2 (*C*H(CH$_3$)$_2$), 27.8 (C(*C*H$_3$)$_3$), 26.5 (C(*C*H$_3$)$_3$), 25.6 (Ph(CH)*C*H$_3$), 25.2 (CH(*C*H$_3$)$_2$), 25.1 (CH(*C*H$_3$)$_2$).

$^{31}P\{^1H\}$ NMR (C_6D_6, 162 MHz): δ [ppm] = 50.6

$^{29}Si\{^1H\}$ NMR (C_6D_6, 79 MHz): δ [ppm] = -83.5 (d, $^2J_{PSi}$ = 48.1 Hz).

IR (ATR): $\tilde{\nu}$ [cm^{-1}] = 3051 (vw), 3031 (vw), 3001 (vw), 2974 (m), 2959 (m), 2929 (w), 2904 (m), 2863 (m), 1583 (m), 1523 (vs), 1500 (s), 1488 (s), 1471 (s), 1455 (s), 1445 (m), 1433 (m), 1392 (s), 1364 (m), 1304 (s), 1258 (m), 1225 (s), 1200 (m), 1179 (m), 1158 (m), 1118 (m), 1105 (m), 1088 (m), 1068 (m), 1034 (m), 1021 (m), 999 (m), 942 (m), 930 (m), 903 (s), 869 (vs), 795 (s), 781 (vs), 773 (s), 744 (vs), 728 (m), 711 (s), 697 (s), 667 (m), 622 (m), 606 (m), 527 (m), 509 (s), 491 (s), 468 (m), 448 (m), 432 (m), 424 (m).

4.9.2 Synthesis of [{(*R*)-PE^{Dipp}PIA}GeCl] (37)

Following the similar procedure described above for **36**, the reaction of [{(*R*)-PE^{Dipp}PIA}Li]$_2$ (300.0 mg, 0.31 mmol, 0.5 eq.) and $GeCl_2$.dioxane (76.0 mg, 0.62 mmol, 1.00 eq.) afforded single crystals from hot heptane suitable for X-ray analysis. The solvent was decanted, and the product was washed with cold *n*-pentane (5 mL).

Yield (based on crystals): 410 mg (40%). **Elemental analysis** calcd. (%) for [$C_{32}H_{36}ClGeN_2P$] (587.71): C 65.40, H 6.17, N 4.77; found: C 65.66, H 6.20, N 4.95.

NMR: Due to decomposition of complex **37** in solution no reasonable spectra could be obtained.

IR (ATR): $\tilde{\nu}$ [cm^{-1}] = 3058 (vw), 3027 (vw), 2973 (m), 2957 (m), 2924 (m), 2865 (m), 2705 (m), 1589 (m), 1493 (m), 1463 (s), 1436 (m), 1382 (m), 1361 (m), 1326 (m), 1273 (m), 1252 (m), 1203 (m), 1189 (m), 1182 (m), 1166 (s), 1117 (s), 1104 (m), 1030 (m), 992 (m), 979 (m), 954 (m), 935 (m), 850 (m), 836 (s), 797 (vs), 783 (m), 752 (m), 722 (m), 702 (m), 694 (m), 667 (s), 622 (vs), 590 (vs), 569 (s), 546 (m), 522 (s), 518 (vs), 491 (m), 458 (m), 433 (m).

4.9.3 Synthesis of [{PhC(N^{t}Bu)$_2$}{(*S*)-PEBA}Si] (38)

Following the similar procedure described above for **36**, the reaction of [{(*S*)-PEBA}Li] (416.0 mg, 1.12 mmol, 1.00 eq.) and [(PhC(N^{t}Bu)$_2$)SiCl] (331.0 mg, 1.12 mmol, 1.00 eq.) afforded single crystals from hot heptane suitable for X-ray analysis. The solvent was decanted, and the product was washed with cold *n*-pentane (5 mL).

Yield (based on crystals): 450 mg (68.3 %). **Elemental analysis** calcd. (%) for [$C_{38}H_{46}SiN_4$] (587.91): C 77.77, H 7.90, N 9.55; found: C 76.98, H 7.61, N 9.55.

^{1}H NMR (C_6D_6, 400 MHz): δ[ppm] = 7.96-7.94 (m, 2 H, Ar-C*H*), 7.56-7.38 (m, 6 H, Ar-C*H*), 7.22-7.15 (m, 7 H, Ar-C*H*), 7.08-7.04 (m, 1 H, Ar-C*H*), 6.95-6.93 (m, 2 H, Ar-C*H*), 6.81 (m, 2 H, Ar-C*H*), 4.82 (q, $^3J_{HH}$ = 6.83 Hz, 1 H, Ph(C*H*)CH$_3$), 4.43 (q, $^3J_{HH}$ = 6.42 Hz, 1 H, Ph(C*H*)CH$_3$), 2.41 (d, $^3J_{HH}$ = 6.87 Hz, 3 H, Ph(CH)C*H*$_3$), 1.34 (d, $^3J_{HH}$ = 6.38 Hz, 3 H, Ph(CH)C*H*$_3$), 1.09 (s, 9H, C(C*H*$_3$)$_3$), 0.90 (s, 9H, C(C*H*$_3$)$_3$).

^{13}C{^{1}H} NMR (C_6D_6, 100 MHz): δ[ppm] = 162.3 (^{tBu}N*C*N), 161.1 (N*C*N), 149.5 (Ar-*C*$_q$), 147.9 (Ar-*C*$_q$), 138.8 (Ar-*C*H), 134.0 (Ar-*C*H), 133.1 (Ar-*C*H), 130.7 (Ar-*C*H), 129.9 (Ar-*C*H), 129.6 (Ar-*C*H), 128.8 (Ar-*C*H), 128.4 (Ar-*C*H), 127.9 (Ar-*C*H), 127.6 (Ar-*C*H), 127.4 (Ar-*C*H), 127.2 (Ar-*C*H), 126.1 (Ar-*C*H), 125.5 (Ar-*C*H), 59.9 (*C*(CH$_3$)$_3$), 53.8 (*C*(CH$_3$)$_3$), 53.4 (Ph(*C*H)CH$_3$), 53.1 (Ph(*C*H)CH$_3$), 31.6 (C(*C*H$_3$)$_3$), 31.4 (C(*C*H$_3$)$_3$), 26.7 (Ph(CH)*C*H$_3$), 23.7 (Ph(CH)*C*H$_3$).

^{29}Si{^{1}H} NMR (C_6D_6, 79 MHz): δ[ppm] = -19.1.

IR (ATR): $\tilde{\nu}$ [cm^{-1}] = 3059 (vw), 3027 (m), 2984 (m), 2962 (m), 2930 (s), 2916 (m), 2866 (m), 1640 (m), 1609 (vs), 1593 (s), 1578 (s), 1492 (m), 1475 (m), 1443 (s), 1400 (vs), 1390 (m), 1361 (m), 1330 (s), 1310 (s), 1298 (vs), 1265 (vs), 1205 (m), 1155 (m), 1124 (vs), 1099 (w), 1073 (m), 1058 (m), 1031 (s), 1017 (m), 1001 (s), 972 (m), 905 (m), 841 (s), 792 (s), 761 (vs), 750 (s), 737 (m), 721 (m), 696 (m), 628 (m), 615 (m), 539 (m), 494 (m).

4.9.4 Synthesis of [{(*S*)-PEBA}GeCl] (39)

Following the similar procedure described above for **36**, the reaction of [{(*S*)-PEBA}Li] (205.0 mg, 0.55 mmol, 1.00 eq.) and $GeCl_2$.dioxane (128.2 mg, 0.55 mmol, 1.00 eq.) afforded single crystals from hot heptane suitable for X-ray analysis. The solvent was decanted, and the product was washed with cold *n*-pentane (5 mL).

Yield (based on crystals): 140 mg (58.3 %). **Elemental analysis** calcd. (%) for [$C_{23}H_{23}ClGeN_2$] (436.54): C 63.43, H 5.32, N 6.43; found: C 63.70, H 5.46, N 6.45

^{1}H NMR (C_6D_6, 400 MHz): δ[ppm] = 7.39-7.37 (m, 2 H, *o*-Ar-C*H*), 7.14-7.12 (m, 3 H, Ar-C*H*), 7.10-7.07 (m, 3 H, Ar-C*H*), 7.04-6.99 (m, 2 H, Ar-C*H*), 6.92-6.89 (m, 1 H, Ar-C*H*), 6.83-6.79 (m, 2 H, Ar-C*H*), 6.68-6.66 (m, 2 H, Ar-C*H*), 4.35 (q, $^3J_{HH}$ = 6.76 Hz, 1 H, Ph(C*H*)CH$_3$), 4.22 (q, $^3J_{HH}$ =

6.77 Hz, 1 H, Ph(C*H*)CH_3), 1.45 (d, $^3J_{HH}$ = 6.77 Hz, 3 H, Ph(CH)CH_3), 1.23 (d, $^3J_{HH}$ = 6.79 Hz, 3 H, Ph(CH)CH_3).

^{13}C{^{1}H} NMR (C_6D_6, 100 MHz): δ [ppm] = 173.2 (NCN), 146.1 (Ar-C_q), 145.7 (Ar-C_q), 130.3 (Ar-*C*H), 129.4 (Ar-*C*H), 129.1 (Ar-*C*H), 128.8 (Ar-*C*H), 128.6 (Ar-*C*H), 127.5 (Ar-*C*H), 127.4 (Ar-*C*H), 127.0 (Ar-*C*H), 126.8 (Ar-*C*H), 126.6 (Ar-*C*H), 56.6 (Ph(*C*H)CH_3), 56.5 (Ph(*C*H)CH_3), 25.8 (Ph(CH)*C*H_3), 24.3 (Ph(CH)*C*H_3).

IR (ATR): $\tilde{\nu}$ [cm^{-1}] = 3058 (vw), 3027 (vw), 2974 (w), 2925 (w), 2900 (vw), 2866 (vw), 1636 (m), 1620 (m), 1601 (m), 1577 (m), 1521 (s), 1493 (w), 1461 (vw), 1449 (vw), 1437 (vw), 1372 (vw), 1348 (m), 1323 (w), 1307 (w), 1278 (w), 1244 (m), 1204 (m), 1181 (m), 1158 (m), 1133 (m), 1104 (m), 1084 (m), 1026 (m), 983 (w), 925 (w), 911 (w), 810 (s), 775 (m), 761 (m), 748 (s), 698 (vs), 648 (w), 583 (m), 552 (vs), 518 (s), 477 (m), 430 (vw).

4.10 Experimental for hydroboration reactions

The catalyst, ferrocene (internal standard) and solid substrates 0.25 mmol (ketone) (if any) were weighed into an NMR tube in the argon filled glove box. C_6D_6 (about 0.5 mL) was condensed into the NMR tube and the mixture was frozen at −196 °C. The reactants 0.3 mmol HBpin and 0.25 mmol ketone (liquid) were injected onto the solid mixture under nitrogen using an oven dried pasture pipette and the whole sample was melted and mixed just before insertion into the core of the NMR machine (t_0). The completion of the reaction was indicated by conversion of singlet of the methyl protons of the ketone into doublet of the final product.

4.10.1 NMR of hydroboration product (^{1}H, ^{11}B, ^{13}C{^{1}H})

NMR data of catalysis: The spectroscopic data for products is in consistence with the corresponding literature.

Ph(Me)CHOBpin[179]

^{1}H NMR (C_6D_6, 400 MHz): δ [ppm] = 7.34 (d, $^3J_{HH}$ = 6.47 Hz, 2 H, *o*-Ar-*H*), 7.15-7.12 (m, 2 H, *m*-Ar-*H*), 7.04 (t, $^3J_{HH}$ = 7.34 Hz, 1 H, *p*-Ar-*H*), 5.38 (q, $^3J_{HH}$ = 6.46 Hz, 1H, PhC*H*CH_3), 1.43 (d, $^3J_{HH}$ = 6.47 Hz, 3H, PhCHCH_3), 1.02 (s, 6H, Bpin-CH_3), 0.99 (s, 6H, Bpin-CH_3).

^{13}C{^{1}H} NMR (C_6D_6, 100 MHz): δ [ppm] = 145.4 (Ar-C_q), 128.5 (Ar-*C*H), 127.4 (Ar-*C*H), 125.7 (Ar-*C*H), 82.5 (Bpin-*C*), 72.9 (O*C*HCH_3Ph), 25.8 (PhCH*C*H_3), 24.7 (Bpin-*C*H_3).

^{11}B NMR (C_6D_6, 128 MHz): δ [ppm] = 22.6.

(4-BrPh)(Me)CHOBpin[180]

^{1}H NMR (C_6D_6, 400 MHz): δ [ppm] = 7.23 (d, $^{3}J_{HH}$ = 8.46 Hz,2 H,*o*-Ar-*H*), 7.00 (d, $^{3}J_{HH}$ = 8.32 Hz,2 H, *m*-Ar-*H*), 5.20 (q, $^{3}J_{HH}$ = 6.44 Hz, 1 H, OC*H*CH_3), 1.32 (d, $^{3}J_{HH}$ = 6.47 Hz, 3 H, OCHCH_3), 1.02 (s, 6 H, Bpin-CH_3), 0.99 (s, 6 H, Bpin-CH_3).

^{13}C{^{1}H} NMR (C_6D_6, 100 MHz): δ [ppm] = 144.3 (Ar-C_q), 131.6 (Ar-*C*H), 127.5 (Ar-*C*H), 121.2 (Ar-*C*H), 82.7 (Bpin-*C*), 72.2 (O*C*HCH_3), 25.5 (OCH*C*H_3), 24.7 (Bpin-*C*H_3).

^{11}B NMR (C_6D_6, 128 MHz): δ [ppm] = 22.4.

(4-NO_2Ph)(Me)CHOBpin[179]

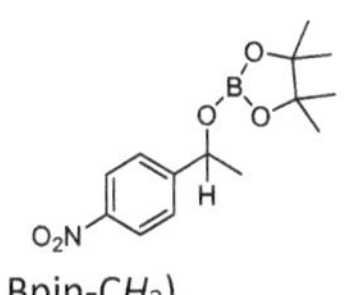

^{1}H NMR (C_6D_6, 400 MHz): δ [ppm] = 7.83 (d, $^{3}J_{HH}$ = 8.78 Hz, 2 H, *o*-Ar-*H*), 7.05 (d, $^{3}J_{HH}$ = 8.52 Hz, 2 H, *m*-Ar-*H*), 5.19 (q, $^{3}J_{HH}$ = 6.48 Hz, 1 H, OC*H*CH3), 1.27 (d, $^{3}J_{HH}$ = 6.51 Hz, 3 H, OCHCH_3), 1.05 (s, 6 H, Bpin-CH_3), 1.02 (s, 6 H, Bpin-CH_3).

^{13}C{^{1}H} NMR (C_6D_6, 100 MHz): δ [ppm] = 152.1 (Ar-C_q), 147.4 (Ar-*C*H), 126.2 (Ar-*C*H), 123.6 (Ar-*C*H), 83.0 (Bpin-*C*), 71.9 (O*C*HCH_3), 25.3 (OCH*C*H_3), 24.6 (Bpin-*C*H_3).

^{11}B NMR (C_6D_6, 128 MHz): δ [ppm] = 22.4.

(4-NH_2Ph)(Me)CHOBpin[181]

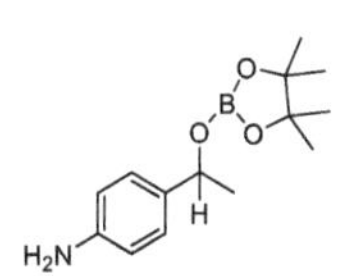

^{1}H NMR (C_6D_6, 400 MHz): δ [ppm] = 7.19 (d, $^{3}J_{HH}$ = 8.20 Hz,2 H, *o*-Ar-*H*), 6.36 (d, $^{3}J_{HH}$ = 8.23 Hz, 2 H, *m*-Ar-*H*), 5.35 (q, $^{3}J_{HH}$ = 6.31 Hz, 1 H, PhC*H*CH_3), 3.05 (s, 2 H, PhNH_2), 1.47 (d, $^{3}J_{HH}$ = 6.40 Hz, 3 H, OCHCH_3), 1.03 (s, 6 H, Bpin-CH_3), 1.01 (s, 6 H, Bpin-CH_3).

^{13}C{^{1}H} NMR (C_6D_6, 100 MHz): δ [ppm] = 146.6 (Ar-C_q), 134.6 (Ar-*C*H), 126.9 (Ar-*C*H), 114.9 (Ar-*C*H), 82.4 (Bpin-*C*), 72.9 (O*C*HCH_3), 25.6 (OCH*C*H_3), 24.7 (Bpin-*C*H_3).

^{11}B NMR (C_6D_6, 128 MHz): δ [ppm] = 22.6.

(4-MePh)(Me)CHOBpin[182]

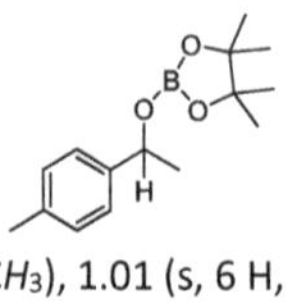

^{1}H NMR (C_6D_6, 400 MHz): δ [ppm] = 7.29 (d, $^3J_{HH}$ = 8.07 Hz, 2 H, *o*-Ar-*H*), 6.97 (d, $^3J_{HH}$ = 7.83 Hz,2 H, *m*-Ar-*H*), 5.41 (q, $^3J_{HH}$ = 6.44 Hz, 1 H, OC*H*CH$_3$), 2.09 (s, 3 H, PhC*H*$_3$), 1.47 (d, $^3J_{HH}$ = 6.46 Hz, 3 H, OCHC*H*$_3$), 1.03 (s, 6 H, Bpin-C*H*$_3$), 1.01 (s, 6 H, Bpin-C*H*$_3$).

^{13}C{^{1}H} NMR (C_6D_6, 100 MHz): δ [ppm] = 142.5 (Ar-C_q), 136.6 (Ar-*C*H), 129.2 (Ar-*C*H), 125.7 (Ar-*C*H), 82.5 (Bpin-*C*), 72.8 (O*C*HCH$_3$), 25.8 (OCH*C*H$_3$), 24.7 (Bpin-*C*H$_3$), 21.1 (Ph*C*H$_3$).

^{11}B NMR (C_6D_6, 128 MHz): δ [ppm] = 22.5.

(4-OCH$_3$Ph)(Me)CHOBpin[180]

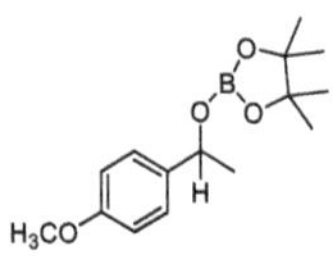

^{1}H NMR (C_6D_6, 400 MHz): δ [ppm] = 7.27 (d, $^3J_{HH}$ = 8.38 Hz, 2 H, *o*-Ar-*H*), 6.75 (d, $^3J_{HH}$ = 8.77 Hz, 2 H, *m*-Ar-*H*), 5.36 (q, $^3J_{HH}$ = 6.42 Hz, 1 H, PhC*H*CH$_3$), 3.33 (s, 3 H, PhOC*H*$_3$), 1.46 (d, $^3J_{HH}$ = 6.44 Hz, 3 H, OCHC*H*$_3$), 1.04 (s, 6 H, Bpin-C*H*$_3$), 1.02 (s, 6 H, Bpin-C*H*$_3$).

^{13}C{^{1}H} NMR (C_6D_6, 100 MHz): δ [ppm] = 159.4 (Ar-C_q), 137.5 (Ar-*C*H), 127.0 (Ar-*C*H), 114.0 (Ar-*C*H), 82.5 (Bpin-*C*), 72.6 (O*C*HCH$_3$Ph), 54.8 (PhO*C*H$_3$), 25.7 (OCH*C*H$_3$)), 25.0 (Bpin-*C*H$_3$).

^{11}B NMR (C_6D_6, 128 MHz): δ [ppm] = 22.5.

(2-ClEt)(Me)CHOBpin[92d]

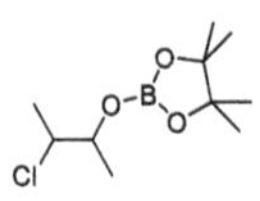

^{1}H NMR (C_6D_6, 400 MHz): δ [ppm] = 4.36-4.29 (m, 1 H, ClC*H*(CH$_3$)), 3.80-3.68 (m, 1 H, (C*H*)CH$_3$(BPin)), 1.30-1.21 (m, 3 H, ClCH(C*H*$_3$)), 1.16-1.12 (m, 3 H, OCHC*H*$_3$), 1.05 (s, 12 H, BPin-C*H*$_3$).

^{13}C{^{1}H} NMR (C_6D_6, 100 MHz): δ [ppm] = 82.7 (BPin-*C*), 74.5 (BpinO*C*CH$_3$), 61.3 (*C*ClCH$_3$), 24.6 (BPin-*C*H$_3$), 20.0 (OCH*C*H$_3$), 19.4 (CCl*C*H$_3$).

^{11}B NMR (C_6D_6, 128 MHz): δ [ppm] = 22.4.

5 Crystal Structure Measurements

5.1 Data Collection and Refinement

A suitable crystal was covered in mineral oil (Aldrich) and mounted on a glass fiber. The crystal was transferred directly to the cold stream of a STOE IPDS 2 or a STOE StadiVari diffractometer. All structures were solved by using the program SHELXS/T[183] and Olex2[184]. The remaining non-hydrogen atoms were located from successive difference Fourier map calculations. The refinements were carried out by using full-matrix least-squares techniques on F^2 by using the program SHELXL.[183] In each case, the locations of the largest peaks in the final difference Fourier map calculations, as well as the magnitude of the residual electron densities, were of no chemical significance.

Refinement Details

The crystal structure of complex **5** contains a molecule of heptane, complex **27** contains half molecule of Et_2O and complex **31**, **35** contain one molecule of toluene in the asymmetric unit. The aforementioned solvent molecule could not be modeled satisfactorily and were therefore squeezed from the electron density map using the OLEX^2 solvent mask.[184] Complex **33** contain a residual peak near the methyl carbon of toluene molecule possibly arising due to symmetry related disorder therefore could not be assigned properly.

5.2 Crystal Data

5.2.1 [(*R*)-HPE^{Dipp}PIA]

Compound	**[(*R*)-HPE^{Dipp}PIA]**
Formula	$C_{32}H_{37}N_2P$
$D_{calc.}$/ g cm^{-3}	1.146
μ/mm^{-1}	0.121
Formula Weight	480.60
Colour	clear colourless
Shape	needle
T/K	100
Crystal System	orthorhombic
Flack Parameter	-0.12(14)
Hooft Parameter	-0.09(8)
Space Group	$P2_12_12_1$
a/Å	8.6016(11)
b/Å	17.371(2)
c/Å	18.640(3)
α/°	
β/°	
γ/°	
V/Å^3	2785.1(6)
Z	4
Z'	1
Wavelength/Å	0.71073
Radiation type	MoK_α
Θ_{min}/°	1.602
Θ_{max}/°	30.161
Measured Refl.	14291
Independent Refl.	7000
Reflections with I > 2(I)	4022
R_{int}	0.0668
Largest Peak	0.330
Deepest Hole	-0.318
GooF	0.991
wR_2 (all data)	0.1685
wR_2	0.1406
R_1 (all data)	0.1364
R_1	0.0686

5.2.2 [(*R*)-HNEPIA]

Compound	[(*R*)-HNEPIA]
Formula	$C_{36}H_{33}N_2P$
$D_{calc.}$/ g cm^{-3}	1.194
μ/mm^{-1}	0.121
Formula Weight	524.61
Colour	colourless
Shape	block
T/K	210
Crystal System	monoclinic
Flack Parameter	-0.01(13)
Hooft Parameter	-0.54(3)
Space Group	$P2_1$
a/Å	17.290(4)
b/Å	9.2600(19)
c/Å	18.360(4)
α/°	
β/°	97.00(3)
γ/°	
V/Å^3	2917.6(10)
Z	4
Z'	2
Wavelength/Å	0.71073
Radiation type	MoK$_\alpha$
Θ_{min}/°	1.528
Θ_{max}/°	29.312
Measured Refl.	31153
Independent Refl.	15184
Reflections with I > 2(I)	8503
R_{int}	0.0607
Largest Peak	0.208
Deepest Hole	-0.273
GooF	1.014
wR_2 (all data)	0.2066
wR_2	0.1737
R_1 (all data)	0.1220
R_1	0.0667

5.2.3 [{(*R*)-PEPIA}$_2$K$_2$] (1)

Compound	**1**
Formula	$C_{56}H_{56}K_2N_4P_2$
$D_{calc.}$/ g cm^{-3}	1.229
μ/mm^{-1}	0.294
Formula Weight	925.18
Colour	colourless
Shape	fragment
T/K	100
Crystal System	orthorhombic
Flack Parameter	-0.032(17)
Hooft Parameter	-0.005(9)
Space Group	$P2_12_12_1$
a/Å	12.481(3)
b/Å	13.110(3)
c/Å	30.547(6)
α/°	
β/°	
γ/°	
V/Å^3	4998.3(17)
Z	4
Z'	1
Wavelength/Å	0.71073
Radiation type	MoK$_\alpha$
Θ_{min}/°	1.690
Θ_{max}/°	26.051
Measured Refl.	41667
Independent Refl.	9830
Reflections with I > 2(I)	9252
R_{int}	0.0372
Largest Peak	0.227
Deepest Hole	-0.250
GooF	1.044
wR_2 (all data)	0.0802
wR_2	0.0789
R_1 (all data)	0.0332
R_1	0.0305

5.2.4 [{(*R*)-PEPIA}$_2$Rb$_2$] (**2**)

Compound	**2**
Formula	$C_{56}H_{56}N_4P_2Rb_2$
$D_{calc.}$/ g cm^{-3}	1.329
μ/mm^{-1}	2.026
Formula Weight	1017.92
Colour	colourless
Shape	prism
T/K	150
Crystal System	orthorhombic
Flack Parameter	-0.011(3)
Hooft Parameter	0.005(3)
Space Group	$P2_12_12_1$
a/Å	12.359(3)
b/Å	13.352(3)
c/Å	30.833(6)
α/°	
β/°	
γ/°	
V/Å^3	5088.1(18)
Z	4
Z'	1
Wavelength/Å	0.71073
Radiation type	MoK$_\alpha$
Θ_{min}/°	1.662
Θ_{max}/°	25.096
Measured Refl.	16064
Independent Refl.	9023
Reflections with I > 2(I)	7626
R_{int}	0.0291
Largest Peak	0.337
Deepest Hole	-0.208
GooF	1.021
wR_2 (all data)	0.0604
wR_2	0.0573
R_1 (all data)	0.0439
R_1	0.0306

5.2.5 [{(*R*)-PEPIA}$_2$Cs$_2$]/[{(*R*)-PEPIA}Cs]$_n$ (**3**)

Compound	**3**
Formula	$C_{84}H_{84}N_6P_3Cs_2$
$D_{calc.}$/ g cm^{-3}	1.447
μ/mm^{-1}	1.530
Formula Weight	1669.21
Colour	colourless
Shape	block
T/K	100
Crystal System	orthorhombic
Flack Parameter	-0.014(11)
Hooft Parameter	-0.018(4)
Space Group	$P2_12_12_1$
a/Å	13.693(3)
b/Å	17.008(3)
c/Å	32.906(7)
α/°	
β/°	
γ/°	
V/Å^3	7663(3)
Z	4
Z'	1
Wavelength/Å	0.71073
Radiation type	MoK$_\alpha$
Θ_{min}/°	2.276
Θ_{max}/°	25.995
Measured Refl.	42195
Independent Refl.	14831
Reflections with I > 2(I)	13858
R_{int}	0.0265
Largest Peak	0.914
Deepest Hole	-1.901
GooF	1.040
wR_2 (all data)	0.1130
wR_2	0.1105
R_1 (all data)	0.0450
R_1	0.0413

5.2.6 [{(*R*)-PE^{Dipp}PIA}$_2$Li$_2$] (**4**)

Compound	**4**
Formula	$C_{64}H_{72}Li_2N_4P_2$
$D_{calc.}$/ g cm^{-3}	1.134
μ/mm^{-1}	0.118
Formula Weight	973.07
Colour	colourless
Shape	prism
T/K	150
Crystal System	orthorhombic
Flack Parameter	-0.15(7)
Hooft Parameter	-0.17(6)
Space Group	$P2_12_12_1$
a/Å	15.640(3)
b/Å	16.860(3)
c/Å	21.620(4)
α/°	
β/°	
γ/°	
V/Å^3	5701(2)
Z	4
Z'	1
Wavelength/Å	0.71073
Radiation type	MoK$_\alpha$
Θ_{min}/°	1.532
Θ_{max}/°	29.247
Measured Refl.	32508
Independent Refl.	15362
Reflections with I > 2(I)	6793
R_{int}	0.0927
Largest Peak	0.290
Deepest Hole	-0.198
GooF	0.704
wR_2 (all data)	0.1611
wR_2	0.1209
R_1 (all data)	0.1555
R_1	0.0552

5.2.7 [{(*R*)-PE^{Dipp}PIA}$_2$Na$_2$] (**5**)

Compound	**5**
Formula	$C_{32}H_{36}N_2NaP$
$D_{calc.}$/ g cm^{-3}	1.089
μ/mm^{-1}	0.125
Formula Weight	502.08
Colour	colourless
Shape	prism
T/K	100
Crystal System	orthorhombic
Flack Parameter	-0.06(4)
Hooft Parameter	-0.07(3)
Space Group	$P2_12_12$
a/Å	20.880(4)
b/Å	22.460(5)
c/Å	13.060(3)
α/°	
β/°	
γ/°	
V/Å^3	6125(2)
Z	8
Z'	2
Wavelength/Å	0.71073
Radiation type	MoK_α
Θ_{min}/°	1.332
Θ_{max}/°	30.316
Measured Refl.	35256
Independent Refl.	15382
Reflections with I > 2(I)	12117
R_{int}	0.0324
Largest Peak	2.584
Deepest Hole	-0.426
GooF	0.939
wR_2 (all data)	0.1803
wR_2	0.1668
R_1 (all data)	0.0791
R_1	0.0591

5.2.8 [{(*R*)-NEPIA}Li(thf)$_2$] (**6**)

Compound	**6**
Formula	$C_{44}H_{48}LiN_2O_2P$
$D_{calc.}$/ g cm^{-3}	1.198
μ/mm^{-1}	0.113
Formula Weight	674.75
Colour	colourless
Shape	plate
T/K	150
Crystal System	monoclinic
Flack Parameter	-0.09(4)
Hooft Parameter	-0.178(16)
Space Group	$P2_1$
a/Å	10.520(2)
b/Å	15.650(3)
c/Å	11.360(2)
α/°	
β/°	90.40(3)
γ/°	
V/Å^3	1870.2(6)
Z	2
Z'	1
Wavelength/Å	0.71073
Radiation type	MoK_α
Θ_{min}/°	1.793
Θ_{max}/°	29.158
Measured Refl.	14108
Independent Refl.	9964
Reflections with I > 2(I)	8390
R_{int}	0.0199
Largest Peak	0.535
Deepest Hole	-0.241
GooF	0.729
wR_2 (all data)	0.1324
wR_2	0.1262
R_1 (all data)	0.0545
R_1	0.0465

5.2.9 [{(*R*)-NEPIA}$_2$Na$_2$] (7)

Compound	**7**
Formula	$C_{72}H_{64}N_4Na_2P_2$
$D_{calc.}$/ g cm^{-3}	1.263
μ/mm^{-1}	0.139
Formula Weight	1093.19
Colour	colourless
Shape	prism
T/K	100
Crystal System	monoclinic
Flack Parameter	-0.07(4)
Hooft Parameter	-0.05(3)
Space Group	$P2_1$
a/Å	13.210(3)
b/Å	15.940(3)
c/Å	13.740(3)
α/°	
β/°	96.60(3)
γ/°	
V/Å^3	2874.0(10)
Z	2
Z'	1
Wavelength/Å	0.71073
Radiation type	MoK$_\alpha$
Θ_{min}/°	1.492
Θ_{max}/°	30.147
Measured Refl.	27640
Independent Refl.	14257
Reflections with I > 2(I)	12622
R_{int}	0.0252
Largest Peak	0.308
Deepest Hole	-0.212
GooF	0.761
wR_2 (all data)	0.1034
wR_2	0.0963
R_1 (all data)	0.0433
R_1	0.0361

5.2.10 [{(*R*)-NEPIA}$_2$K$_2$] (**8**)

Compound	**8**
Formula	$C_{36}H_{32}KN_2P$
$D_{calc.}$/ g cm^{-3}	1.283
μ/mm^{-1}	0.266
Formula Weight	562.70
Colour	colourless
Shape	prism
T/K	150
Crystal System	monoclinic
Flack Parameter	0.02(4)
Hooft Parameter	0.097(16)
Space Group	$P2_1$
a/Å	13.050(3)
b/Å	15.560(3)
c/Å	14.440(3)
α/°	
β/°	96.70(3)
γ/°	
V/Å^3	2912.1(10)
Z	4
Z'	2
Wavelength/Å	0.71073
Radiation type	MoK_α
Θ_{min}/°	1.571
Θ_{max}/°	25.999
Measured Refl.	16286
Independent Refl.	11259
Reflections with I > 2(I)	8795
R_{int}	0.0413
Largest Peak	0.234
Deepest Hole	-0.357
GooF	1.014
wR_2 (all data)	0.0943
wR_2	0.0858
R_1 (all data)	0.0699
R_1	0.0439

5.2.11 [{(*R*)-PEPIA}$_2$Ca] (9)

Compound	**9**
Formula	$C_{56}H_{56}CaN_4P_2$
$D_{calc.}$/ g cm^{-3}	1.211
μ/mm^{-1}	0.236
Formula Weight	887.06
Colour	clear colourless
Shape	Prism
T/K	150
Crystal System	monoclinic
Flack Parameter	-0.06(3)
Hooft Parameter	-0.077(9)
Space Group	*C*2
a/Å	21.970(4)
b/Å	13.520(3)
c/Å	18.050(4)
α/°	
β/°	114.80(3)
γ/°	
V/Å^3	4867(2)
Z	4
Z'	1
Wavelength/Å	0.71073
Radiation type	MoK$_\alpha$
Θ_{min}/°	1.820
Θ_{max}/°	29.491
Measured Refl.	20615
Independent Refl.	13402
Reflections with I > 2(I)	10867
R_{int}	0.0406
Largest Peak	0.424
Deepest Hole	-0.329
GooF	1.046
wR_2 (all data)	0.1220
wR_2	0.1170
R_1 (all data)	0.0608
R_1	0.0475

5.2.12 [{(*R*)-PE^{Dipp}PIA}Mg{N(SiMe$_3$)$_2$}(thf)] (10)

Compound	**10**
Formula	$C_{42}H_{62}MgN_3OPSi_2$
$D_{calc.}$/ g cm^{-3}	1.156
μ/mm^{-1}	0.171
Formula Weight	736.40
Colour	Colourless
Shape	Prism
T/K	100
Crystal System	orthorhombic
Flack Parameter	0.02(4)
Hooft Parameter	0.03(3)
Space Group	$P2_12_12_1$
a/Å	14.030(3)
b/Å	16.690(3)
c/Å	18.070(4)
α/°	
β/°	
γ/°	
V/Å^3	4231.3(15)
Z	4
Z'	1
Wavelength/Å	0.71073
Radiation type	MoK_α
Θ_{min}/°	1.661
Θ_{max}/°	30.284
Measured Refl.	48457
Independent Refl.	11347
Reflections with I > 2(I)	9220
R_{int}	0.0387
Largest Peak	0.316
Deepest Hole	-0.201
GooF	0.711
wR_2 (all data)	0.1033
wR_2	0.0954
R_1 (all data)	0.0520
R_1	0.0372

5.2.13 [{(*R*)-PE^{Dipp}PIA}$_2$Mg] (11)

Compound	**11**
Formula	$C_{64}H_{72}MgN_4P_2$
$D_{calc.}$/ g cm^{-3}	1.149
μ/mm^{-1}	0.130
Formula Weight	983.50
Colour	colourless
Shape	prism
T/K	210
Crystal System	orthorhombic
Flack Parameter	-0.07(5)
Hooft Parameter	-0.12(3)
Space Group	$P2_12_12_1$
a/Å	14.020(3)
b/Å	18.350(4)
c/Å	22.090(4)
α/°	
β/°	
γ/°	
V/Å^3	5683(2)
Z	4
Z'	1
Wavelength/Å	0.71073
Radiation type	MoK_α
Θ_{min}/°	1.443
Θ_{max}/°	29.243
Measured Refl.	49441
Independent Refl.	15332
Reflections with I > 2(I)	10698
R_{int}	0.0562
Largest Peak	0.347
Deepest Hole	-0.197
GooF	0.999
wR_2 (all data)	0.1628
wR_2	0.1379
R_1 (all data)	0.0925
R_1	0.0548

5.2.14 [{(*R*)-PE[Dipp]PIA}$_2$Ca] (12)

Compound	**12**
Formula	$C_{64}H_{72}N_4P_2Ca$
$D_{calc.}$/ g cm^{-3}	1.223
μ/mm^{-1}	0.219
Formula Weight	999.27
Colour	colourless
Shape	block
T/K	100
Crystal System	orthorhombic
Flack Parameter	0.00(5)
Hooft Parameter	-0.03(2)
Space Group	$P2_12_12_1$
a/Å	13.680(3)
b/Å	17.800(4)
c/Å	22.280(5)
α/°	
β/°	
γ/°	
V/Å^3	5425.3(19)
Z	4
Z'	1
Wavelength/Å	0.71073
Radiation type	MoK_α
Θ_{min}/°	1.464
Θ_{max}/°	30.203
Measured Refl.	28230
Independent Refl.	13890
Reflections with I > 2(I)	10819
R_{int}	0.0672
Largest Peak	1.841
Deepest Hole	-0.836
GooF	1.074
wR_2 (all data)	0.2335
wR_2	0.2140
R_1 (all data)	0.1096
R_1	0.0830

5.2.15 [{(*R*)-PE^{Dipp}PIA}$_2$Yb] (13)

Compound	**13**
Formula	$C_{69}H_{84}N_4P_2Yb$
$D_{calc.}$/ g cm^{-3}	1.274
μ/mm^{-1}	1.582
Formula Weight	1204.38
Colour	red
Shape	irregular
T/K	150
Crystal System	monoclinic
Flack Parameter	-0.029(5)
Hooft Parameter	-0.014(3)
Space Group	$P2_1$
a/Å	11.010(2)
b/Å	17.660(4)
c/Å	16.250(3)
α/°	
β/°	96.30(3)
γ/°	
V/Å^3	3140.5(11)
Z	2
Z'	1
Wavelength/Å	0.71073
Radiation type	MoK$_\alpha$
Θ_{min}/°	1.709
Θ_{max}/°	31.543
Measured Refl.	32307
Independent Refl.	16830
Reflections with I > 2(I)	13884
R_{int}	0.0347
Largest Peak	0.924
Deepest Hole	-0.433
GooF	0.972
wR_2 (all data)	0.0771
wR_2	0.0730
R_1 (all data)	0.0518
R_1	0.0368

5.2.16 [{(*R*)-NEPIA}$_2$Ca] (14)

Compound	14
Formula	$C_{72}H_{64}N_4P_2Ca$
$D_{calc.}$/ g cm^{-3}	1.248
μ/mm^{-1}	0.211
Formula Weight	1087.29
Colour	colourless
Shape	prism
T/K	100
Crystal System	monoclinic
Flack Parameter	-0.044(13)
Hooft Parameter	-0.029(9)
Space Group	$P2_1$
a/Å	13.200(3)
b/Å	16.020(3)
c/Å	13.770(3)
α/°	
β/°	96.60(3)
γ/°	
V/Å^3	2892.6(10)
Z	2
Z'	1
Wavelength/Å	0.71073
Radiation type	MoK$_\alpha$
Θ_{min}/°	1.489
Θ_{max}/°	30.232
Measured Refl.	29150
Independent Refl.	14590
Reflections with I > 2(I)	12947
R_{int}	0.0235
Largest Peak	0.330
Deepest Hole	-0.234
GooF	1.005
wR_2 (all data)	0.0831
wR_2	0.0801
R_1 (all data)	0.0428
R_1	0.0351

5.2.17 [{(*R*)-PE^{Dipp}PIA}$_2$Cu$_2$] (15)

Compound	15*toluene
Formula	$C_{71}H_{80}Cu_2N_4P_2$
$D_{calc.}$/ g cm^{-3}	1.236
μ/mm^{-1}	0.766
Formula Weight	1178.41
Colour	colourless
Shape	prism
T/K	210
Crystal System	orthorhombic
Flack Parameter	-0.009(5)
Hooft Parameter	-0.014(3)
Space Group	$P2_12_12_1$
a/Å	13.4500(19)
b/Å	13.4500(19)
c/Å	35.010(7)
α/°	
β/°	
γ/°	
V/Å^3	6333.4(18)
Z	4
Z'	1
Wavelength/Å	0.71073
Radiation type	MoK$_\alpha$
Θ_{min}/°	1.622
Θ_{max}/°	25.166
Measured Refl.	33505
Independent Refl.	11237
Reflections with I > 2(I)	9109
R_{int}	0.0281
Largest Peak	0.411
Deepest Hole	-0.185
GooF	0.963
wR_2 (all data)	0.0886
wR_2	0.0866
R_1 (all data)	0.0460
R_1	0.0364

5.2.18 [{(*R*)-PE^{Dipp}PIA}$_2$Zn] (16)

Compound	16
Formula	$C_{64}H_{72}N_4P_2Zn$
$D_{calc.}$/ g cm^{-3}	1.199
μ/mm^{-1}	0.531
Formula Weight	1024.56
Colour	yellow
Shape	prism
T/K	220
Crystal System	orthorhombic
Flack Parameter	-0.020(5)
Hooft Parameter	-0.009(2)
Space Group	$P2_12_12_1$
a/Å	13.930(3)
b/Å	18.380(4)
c/Å	22.170(4)
α/°	
β/°	
γ/°	
V/Å^3	5676(2)
Z	4
Z'	1
Wavelength/Å	0.71073
Radiation type	MoK_α
Θ_{min}/°	1.439
Θ_{max}/°	25.181
Measured Refl.	28520
Independent Refl.	10101
Reflections with I > 2(I)	8675
R_{int}	0.0216
Largest Peak	0.390
Deepest Hole	-0.154
GooF	0.796
wR_2 (all data)	0.1014
wR_2	0.0915
R_1 (all data)	0.0434
R_1	0.0331

5.2.19 [{(R)-NEPIA}$_2$Cu$_2$] (17)

Compound	**17**
Formula	$C_{72}H_{64}Cu_2N_4P_2$
$D_{calc.}$/ g cm^{-3}	1.360
μ/mm^{-1}	0.846
Formula Weight	1174.29
Colour	colourless
Shape	prism
T/K	100.0
Crystal System	orthorhombic
Flack Parameter	-0.018(8)
Hooft Parameter	-0.004(3)
Space Group	$P2_12_12_1$
a/Å	10.3351(2)
b/Å	19.1071(4)
c/Å	29.0438(7)
α/°	
β/°	
γ/°	
V/Å^3	5735.4(2)
Z	4
Z'	1
Wavelength/Å	0.71073
Radiation type	MoK$_\alpha$
Θ_{min}/°	2.240
Θ_{max}/°	27.032
Measured Refl.	37721
Independent Refl.	11619
Reflections with I > 2(I)	10700
R_{int}	0.0479
Largest Peak	1.272
Deepest Hole	-0.704
GooF	1.064
wR_2 (all data)	0.1382
wR_2	0.1314
R_1 (all data)	0.0556
R_1	0.0492

5.2.20 [{DippPIA}$_2$Cu$_2$] (**18**)

Compound	**18*toluene**
Formula	$C_{79}H_{96}Cu_2N_4P_2$
$D_{calc.}$/ g cm^{-3}	1.264
μ/mm^{-1}	3.902
Formula Weight	1290.61
Colour	Orange
Shape	Block
T/K	150
Crystal System	monoclinic
Space Group	$P2_1/n$
a/Å	13.6050(5)
b/Å	11.7346(3)
c/Å	21.2378(7)
α/°	
β/°	90.668(3)
γ/°	
V/Å^3	3390.37(19)
Z	2
Z'	0.5
Wavelength/Å	1.34143
Radiation type	GaK_α
Θ_{min}/°	3.339
Θ_{max}/°	64.378
Measured Refl.	25537
Independent Refl.	8281
Reflections with I > 2(I)	5404
R_{int}	0.0509
Largest Peak	0.693
Deepest Hole	-0.880
GooF	1.062
wR_2 (all data)	0.2006
wR_2	0.1751
R_1 (all data)	0.0976
R_1	0.0653

5.2.21 [{DippPIA}ZnPh] (19)

Compound	19
Formula	$C_{84}H_{98}N_4P_2Zn_2$
$D_{calc.}$/ g cm^{-3}	1.187
μ/mm^{-1}	0.719
Formula Weight	1356.34
Colour	colourless
Shape	plate
T/K	220
Crystal System	monoclinic
Space Group	$P2_1/n$
a/Å	12.020(2)
b/Å	40.480(8)
c/Å	15.630(3)
α/°	
β/°	93.50(3)
γ/°	
V/Å^3	7591(3)
Z	4
Z'	1
Wavelength/Å	0.71073
Radiation type	MoK_α
Θ_{min}/°	1.399
Θ_{max}/°	25.155
Measured Refl.	35060
Independent Refl.	13506
Reflections with I > 2(I)	10103
R_{int}	0.0272
Largest Peak	0.363
Deepest Hole	-0.283
GooF	1.019
wR_2 (all data)	0.1285
wR_2	0.1162
R_1 (all data)	0.0666
R_1	0.0450

5.2.22 [{(*R*)-PE^{Dipp}PIA}AlMe$_2$] (**20**)

Compound	**20**
Formula	$C_{34}H_{42}AlN_2P$
$D_{calc.}$/ g cm^{-3}	1.140
μ/mm^{-1}	0.140
Formula Weight	536.64
Colour	clear colourless
Shape	Prism
T/K	150
Crystal System	orthorhombic
Flack Parameter	0.00(4)
Hooft Parameter	0.03(2)
Space Group	$P2_12_12_1$
a/Å	9.1016(4)
b/Å	17.9217(13)
c/Å	19.1767(12)
α/°	
β/°	
γ/°	
V/Å^3	3128.0(3)
Z	4
Z'	1
Wavelength/Å	0.71073
Radiation type	MoK_α
Θ_{min}/°	2.124
Θ_{max}/°	29.463
Measured Refl.	16288
Independent Refl.	8652
Reflections with I > 2(I)	6945
R_{int}	0.0237
Largest Peak	0.259
Deepest Hole	-0.207
GooF	0.988
wR_2 (all data)	0.0797
wR_2	0.0765
R_1 (all data)	0.0493
R_1	0.0354

5.2.23 [{(*R*)-PE^{Dipp}PIA}AlMe(C_6F_5)] (**21**)

Compound	**21**
Formula	$C_{78}H_{78}Al_2F_{10}N_4P_2$
$D_{calc.}$/ g cm^{-3}	1.271
μ/mm^{-1}	0.156
Formula Weight	1377.34
Colour	colourless
Shape	Plate
T/K	150
Crystal System	monoclinic
Flack Parameter	-0.03(7)
Hooft Parameter	-0.20(3)
Space Group	$P2_1$
a/Å	18.350(4)
b/Å	9.1700(18)
c/Å	21.450(4)
α/°	
β/°	94.10(3)
γ/°	
V/Å^3	3600.1(13)
Z	2
Z'	1
Wavelength/Å	0.71073
Radiation type	MoK$_\alpha$
Θ_{min}/°	1.515
Θ_{max}/°	29.238
Measured Refl.	34735
Independent Refl.	18651
Reflections with I > 2(I)	12355
R_{int}	0.0368
Largest Peak	0.333
Deepest Hole	-0.204
GooF	0.920
wR_2 (all data)	0.1550
wR_2	0.1435
R_1 (all data)	0.0867
R_1	0.0540

5.2.24 [{(*R*)-PE^{Dipp}PIA}$AlCl_2$] (**22**)

Compound	**22**
Formula	$C_{32}H_{36}AlCl_2N_2P$
$D_{calc.}$/ g cm^{-3}	1.250
μ/mm^{-1}	0.316
Formula Weight	577.48
Colour	colourless
Shape	prism
T/K	100
Crystal System	orthorhombic
Flack Parameter	-0.02(3)
Hooft Parameter	-0.02(2)
Space Group	$P2_12_12_1$
a/Å	17.415(4)
b/Å	17.966(4)
c/Å	9.810(2)
α/°	
β/°	
γ/°	
V/Å^3	3069.3(11)
Z	4
Z'	1
Wavelength/Å	0.71073
Radiation type	MoK_α
Θ_{min}/°	1.629
Θ_{max}/°	31.447
Measured Refl.	18257
Independent Refl.	8409
Reflections with I > 2(I)	7043
R_{int}	0.0260
Largest Peak	0.435
Deepest Hole	-0.239
GooF	0.829
wR_2 (all data)	0.1173
wR_2	0.1079
R_1 (all data)	0.0554
R_1	0.0413

5.2.25 [{(*R*)-PEPIA}$_2$AlCl] (**23**)

Compound	**23*{0.5(Et$_2$)O}**
Formula	$C_{58}H_{61}AlClN_4OP_2$
$D_{calc.}$/ g cm^{-3}	1.160
μ/mm^{-1}	0.186
Formula Weight	954.47
Colour	clear colourless
Shape	Prism
T/K	150
Crystal System	orthorhombic
Flack Parameter	0.05(3)
Hooft Parameter	0.01(2)
Space Group	$P2_12_12$
a/Å	18.3994(4)
b/Å	14.4577(5)
c/Å	10.2758(7)
α/°	
β/°	
γ/°	
V/Å^3	2733.5(2)
Z	2
Z'	0.5
Wavelength/Å	0.71073
Radiation type	MoK$_\alpha$
Θ_{min}/°	1.791
Θ_{max}/°	27.134
Measured Refl.	16111
Independent Refl.	6036
Reflections with I > 2(I)	4707
R_{int}	0.0261
Largest Peak	0.727
Deepest Hole	-0.586
GooF	1.058
wR_2 (all data)	0.1697
wR_2	0.1546
R_1 (all data)	0.0787
R_1	0.0575

5.2.26 [{(*R*)-PEPIA}$_2$GaCl] (**24**)

Compound	**24*{0.5(Et$_2$O)}**
Formula	$C_{58}H_{61}ClGaN_4O_{0.5}P_2$
$D_{calc.}$/ g cm^{-3}	1.188
μ/mm^{-1}	0.644
Formula Weight	989.21
Colour	clear colourless
Shape	Plate
T/K	240
Crystal System	orthorhombic
Flack Parameter	-0.012(15)
Hooft Parameter	-0.047(8)
Space Group	$P2_12_12$
a/Å	18.4844(4)
b/Å	14.5247(8)
c/Å	10.2958(8)
α/°	
β/°	
γ/°	
V/Å^3	2764.2(3)
Z	2
Z'	0.5
Wavelength/Å	0.71073
Radiation type	MoK_α
Θ_{min}/°	1.783
Θ_{max}/°	29.493
Measured Refl.	19503
Independent Refl.	7672
Reflections with I > 2(I)	3984
R_{int}	0.0639
Largest Peak	0.685
Deepest Hole	-0.287
GooF	0.983
wR_2 (all data)	0.1979
wR_2	0.1590
R_1 (all data)	0.1388
R_1	0.0635

5.2.27 [{(*R*)-PEPIA}$_2$Al]$^+$[$GaCl_4$]$^-$ (**25**)

Compound	**25**
Formula	$C_{56}H_{56}AlCl_4GaN_4P_2$
$D_{calc.}$/ g cm^{-3}	1.341
μ/mm^{-1}	0.827
Formula Weight	1085.48
Colour	clear colourless
Shape	irregular
T/K	150
Crystal System	trigonal
Flack Parameter	-0.024(8)
Hooft Parameter	0.003(7)
Space Group	$P3_121$
a/Å	12.6047(3)
b/Å	12.6047(3)
c/Å	58.597(2)
α/°	
β/°	
γ/°	120
V/Å^3	8062.5(5)
Z	6
Z'	1
Wavelength/Å	0.71073
Radiation type	MoK_α
Θ_{min}/°	1.866
Θ_{max}/°	25.553
Measured Refl.	20530
Independent Refl.	9628
Reflections with I > 2(I)	6842
R_{int}	0.0597
Largest Peak	0.569
Deepest Hole	-0.431
GooF	0.942
wR_2 (all data)	0.0813
wR_2	0.0767
R_1 (all data)	0.0771
R_1	0.0485

5.2.28 [{(*R*)-PEPIA}$_2$Ga]$^+$[AlCl$_4$]$^-$ (26)

Compound	**26**
Formula	$C_{56}H_{56}AlCl_4GaN_4P_2$
$D_{calc.}$/ g cm^{-3}	1.331
μ/mm^{-1}	0.821
Formula Weight	1085.48
Colour	clear colourless
Shape	Prism
T/K	150
Crystal System	trigonal
Flack Parameter	0.024(15)
Hooft Parameter	0.019(10)
Space Group	$P3_121$
a/Å	12.6494(2)
b/Å	12.6494(2)
c/Å	58.6453(12)
α/°	
β/°	
γ/°	120
V/Å^3	8126.5(3)
Z	6
Z'	1
Wavelength/Å	0.71073
Radiation type	MoK_α
Θ_{min}/°	1.859
Θ_{max}/°	25.606
Measured Refl.	28278
Independent Refl.	10102
Reflections with I > 2(I)	6721
R_{int}	0.1149
Largest Peak	0.423
Deepest Hole	-0.402
GooF	1.034
wR_2 (all data)	0.1306
wR_2	0.1145
R_1 (all data)	0.1150
R_1	0.0654

5.2.29 [{(*R*)-PEPIA}$_2$AlH] (**27**)

Compound	**27**
Formula	$C_{56}H_{57}AlN_4P_2$
$D_{calc.}$/ g cm^{-3}	1.108
μ/mm^{-1}	0.138
Formula Weight	874.97
Colour	clear colourless
Shape	rhombohedral
T/K	150
Crystal System	orthorhombic
Flack Parameter	-0.06(3)
Hooft Parameter	-0.02(2)
Space Group	$P2_12_12$
a/Å	18.1730(2)
b/Å	14.6537(5)
c/Å	9.8443(6)
α/°	
β/°	
γ/°	
V/Å^3	2621.55(19)
Z	2
Z'	0.5
Wavelength/Å	0.71073
Radiation type	MoK_α
Θ_{min}/°	1.785
Θ_{max}/°	29.546
Measured Refl.	35079
Independent Refl.	7327
Reflections with I > 2(I)	5686
R_{int}	0.0424
Largest Peak	0.317
Deepest Hole	-0.184
GooF	0.947
wR_2 (all data)	0.0862
wR_2	0.0819
R_1 (all data)	0.0554
R_1	0.0384

5.2.30 [{(*R*)-PEPIA}Hf(NMe_2)$_3$] (**28**)

Compound	**28**
Formula	$C_{34}H_{46}HfN_5P$
$D_{calc.}$/ g cm^{-3}	1.463
μ/mm^{-1}	3.208
Formula Weight	734.22
Colour	colourless
Shape	prism
T/K	150
Crystal System	orthorhombic
Flack Parameter	-0.017(5)
Hooft Parameter	-0.013(2)
Space Group	$P2_12_12_1$
a/Å	9.2500(19)
b/Å	13.160(3)
c/Å	27.380(6)
α/°	
β/°	
γ/°	
V/Å^3	3333.0(12)
Z	4
Z'	1
Wavelength/Å	0.71073
Radiation type	MoK$_\alpha$
Θ_{min}/°	1.487
Θ_{max}/°	25.545
Measured Refl.	33111
Independent Refl.	6208
Reflections with I > 2(I)	6014
R_{int}	0.0390
Largest Peak	0.391
Deepest Hole	-0.591
GooF	1.030
wR_2 (all data)	0.0435
wR_2	0.0431
R_1 (all data)	0.0184
R_1	0.0172

5.2.31 [{(*R*)-PE^{Dipp}PIA}Zr(NMe$_2$)$_3$] (29)

Compound	**29**
Formula	$C_{38}H_{54}N_5PZr$
$D_{calc.}$/ g cm^{-3}	1.243
μ/mm^{-1}	0.367
Formula Weight	703.05
Colour	colourless
Shape	Prism
T/K	150
Crystal System	orthorhombic
Flack Parameter	-0.044(14)
Hooft Parameter	-0.033(10)
Space Group	$P2_12_12_1$
a/Å	11.600(2)
b/Å	13.370(3)
c/Å	48.440(10)
α/°	
β/°	
γ/°	
V/Å^3	7513(3)
Z	8
Z'	2
Wavelength/Å	0.71073
Radiation type	MoK_α
Θ_{min}/°	1.580
Θ_{max}/°	25.173
Measured Refl.	40108
Independent Refl.	13303
Reflections with I > 2(I)	9937
R_{int}	0.0367
Largest Peak	0.395
Deepest Hole	-0.208
GooF	0.958
wR_2 (all data)	0.0696
wR_2	0.0675
R_1 (all data)	0.0513
R_1	0.0333

5.2.32 [{(*R*)-PE^{Dipp}PIA}Hf(NMe_2)$_3$] (30)

Compound	**30**
Formula	$C_{76}H_{108}Hf_2N_{10}P_2$
$D_{calc.}$/ g cm^{-3}	1.400
μ/mm^{-1}	2.856
Formula Weight	1580.64
Colour	colourless
Shape	Prism
T/K	150
Crystal System	orthorhombic
Flack Parameter	-0.019(4)
Hooft Parameter	-0.0209(18)
Space Group	$P2_12_12_1$
a/Å	11.600(2)
b/Å	13.360(3)
c/Å	48.400(10)
α/°	
β/°	
γ/°	
V/Å^3	7501(3)
Z	4
Z'	1
Wavelength/Å	0.71073
Radiation type	MoK_α
Θ_{min}/°	1.581
Θ_{max}/°	25.165
Measured Refl.	34420
Independent Refl.	13210
Reflections with I > 2(I)	12160
R_{int}	0.0236
Largest Peak	1.175
Deepest Hole	-0.699
GooF	0.774
wR_2 (all data)	0.0936
wR_2	0.0909
R_1 (all data)	0.0366
R_1	0.0318

5.2.33 [{(*R*)-PEPIA}$_3$Y] (**31**)

Compound	**31**
Formula	$C_{84}H_{84}N_6P_3Y$
$D_{calc.}$/ g cm^{-3}	1.042
μ/mm^{-1}	0.769
Formula Weight	1359.39
Colour	colourless
Shape	block
T/K	100
Crystal System	tetragonal
Flack Parameter	-0.005(7)
Hooft Parameter	0.005(2)
Space Group	$P4_12_12$
a/Å	20.020(3)
b/Å	20.020(3)
c/Å	21.620(4)
α/°	
β/°	
γ/°	
V/Å^3	8665(3)
Z	4
Z'	0.5
Wavelength/Å	0.71073
Radiation type	MoK$_\alpha$
Θ_{min}/°	2.242
Θ_{max}/°	31.370
Measured Refl.	36773
Independent Refl.	11711
Reflections with I > 2(I)	7145
R_{int}	0.0614
Largest Peak	0.644
Deepest Hole	-1.015
GooF	1.388
wR_2 (all data)	0.2359
wR_2	0.2058
R_1 (all data)	0.1451
R_1	0.0792

5.2.34 [{(*R*)-PEPIA}$_3$La] (**32**)

Compound	**32*thf**
Formula	$C_{88}H_{92}LaN_6OP_3$
$D_{calc.}$/ g cm^{-3}	1.316
μ/mm^{-1}	0.688
Formula Weight	1481.49
Colour	colourless
Shape	prism
T/K	100
Crystal System	orthorhombic
Flack Parameter	-0.04(2)
Hooft Parameter	-0.010(11)
Space Group	$P2_12_12_1$
a/Å	25.970(5)
b/Å	14.400(3)
c/Å	19.990(4)
α/°	
β/°	
γ/°	
V/Å^3	7476(3)
Z	4
Z'	1
Wavelength/Å	0.71073
Radiation type	MoK_α
Θ_{min}/°	2.345
Θ_{max}/°	29.687
Measured Refl.	36041
Independent Refl.	17501
Reflections with I > 2(I)	11136
R_{int}	0.0596
Largest Peak	0.914
Deepest Hole	-1.171
GooF	0.777
wR_2 (all data)	0.1476
wR_2	0.1218
R_1 (all data)	0.0964
R_1	0.0522

5.2.35 [{(*R*)-PEPIA}$_3$Tb] (**33**)

Compound	**33*(4·toluene)**
Formula	$C_{196}H_{199}N_{12}P_6Tb_2$
$D_{calc.}$/ g cm^{-3}	1.241
μ/mm^{-1}	0.924
Formula Weight	3226.32
Colour	colourless
Shape	plate
T/K	100
Crystal System	monoclinic
Flack Parameter	-0.012(3)
Hooft Parameter	-0.0138(9)
Space Group	$P2_1$
a/Å	19.620(4)
b/Å	15.350(3)
c/Å	29.120(6)
α/°	
β/°	100.20(3)
γ/°	
V/Å^3	8631(3)
Z	2
Z'	1
Wavelength/Å	0.71073
Radiation type	MoK$_\alpha$
Θ_{min}/°	1.372
Θ_{max}/°	29.534
Measured Refl.	167053
Independent Refl.	44273
Reflections with I > 2(I)	41251
R_{int}	0.0225
Largest Peak	2.709
Deepest Hole	-1.652
GooF	1.837
wR_2 (all data)	0.2179
wR_2	0.2136
R_1 (all data)	0.0609
R_1	0.0556

5.2.36 [{(*R*)-PEPIA}$_3$Yb] (34)

Compound	34*toluene
Formula	$C_{91}H_{92}N_6P_3Yb$
$D_{calc.}$/ g cm^{-3}	1.157
μ/mm^{-1}	1.159
Formula Weight	1535.65
Colour	colourless
Shape	prism
T/K	210
Crystal System	monoclinic
Flack Parameter	-0.003(5)
Hooft Parameter	0.0057(19)
Space Group	$P2_1$
a/Å	19.660(4)
b/Å	15.500(3)
c/Å	29.350(6)
α/°	
β/°	99.70(3)
γ/°	
V/Å^3	8816(3)
Z	4
Z'	2
Wavelength/Å	0.71073
Radiation type	MoK$_\alpha$
Θ_{min}/°	1.490
Θ_{max}/°	25.523
Measured Refl.	50122
Independent Refl.	32642
Reflections with I > 2(I)	27525
R_{int}	0.0299
Largest Peak	7.968
Deepest Hole	-4.338
GooF	1.151
wR_2 (all data)	0.2863
wR_2	0.2809
R_1 (all data)	0.0999
R_1	0.0895

5.2.37 [{(*R*)-PEPIA}$_3$Lu] (**35**)

Compound	**35**
Formula	$C_{84}H_{84}LuN_6P_3$
$D_{calc.}$/ g cm^{-3}	1.0912
μ/mm^{-1}	1.217
Formula Weight	1445.53
Colour	colourless
Shape	prism
T/K	150
Crystal System	tetragonal
Flack Parameter	-0.018(13)
Hooft Parameter	0.0116(19)
Space Group	$P4_12_12$
a/Å	20.030(3)
b/Å	20.030(3)
c/Å	21.930(4)
α/°	
β/°	
γ/°	
V/Å^3	8798(2)
Z	4
Z'	0.5
Wavelength/Å	0.71073
Radiation type	Mo K$_\alpha$
Θ_{min}/°	1.71
Θ_{max}/°	29.27
Measured Refl.	64482
Independent Refl.	11889
Reflections with I > 2(I)	9970
R_{int}	0.0457
Largest Peak	0.5400
Deepest Hole	-1.0207
GooF	1.0147
wR_2 (all data)	0.1079
wR_2	0.1009
R_1 (all data)	0.0475
R_1	0.0380

5.2.38 [{PhC(N^{t}Bu)$_2$}Si{N(*R*)(CH)MePh}{(=NDipp)PPh$_2$}] (**36**)

Compound	**36**
Formula	$C_{47}H_{59}N_4PSi$
$D_{calc.}$/ g cm^{-3}	1.157
μ/mm^{-1}	0.130
Formula Weight	739.04
Colour	yellow
Shape	needle
T/K	100
Crystal System	orthorhombic
Flack Parameter	0.09(11)
Hooft Parameter	0.09(8)
Space Group	$P2_12_12_1$
a/Å	15.250(3)
b/Å	15.470(3)
c/Å	17.990(4)
α/°	
β/°	
γ/°	
V/Å^3	4244.2(15)
Z	4
Z'	1
Wavelength/Å	0.71073
Radiation type	MoK$_\alpha$
Θ_{min}/°	1.736
Θ_{max}/°	31.512
Measured Refl.	38350
Independent Refl.	12141
Reflections with I > 2(I)	5883
R_{int}	0.1192
Largest Peak	0.335
Deepest Hole	-0.386
GooF	0.963
wR_2 (all data)	0.2177
wR_2	0.1621
R_1 (all data)	0.2064
R_1	0.0824

5.2.39 [{(*R*)-PE^{Dipp}PIA}GeCl] (37)

Compound	37
Formula	$C_{32}H_{36}ClGeN_2P$
$D_{calc.}$/ g cm^{-3}	1.310
μ/mm^{-1}	1.195
Formula Weight	587.64
Colour	colourless
Shape	prism
T/K	150
Crystal System	orthorhombic
Flack Parameter	-0.006(4)
Hooft Parameter	0.003(2)
Space Group	$P2_12_12_1$
a/Å	11.450(2)
b/Å	12.300(3)
c/Å	21.150(4)
α/°	
β/°	
γ/°	
V/Å^3	2978.7(10)
Z	4
Z'	1
Wavelength/Å	0.71073
Radiation type	MoK_α
Θ_{min}/°	1.915
Θ_{max}/°	29.489
Measured Refl.	21623
Independent Refl.	8267
Reflections with I > 2(I)	7342
R_{int}	0.0256
Largest Peak	1.461
Deepest Hole	-0.499
GooF	1.037
wR_2 (all data)	0.0903
wR_2	0.0877
R_1 (all data)	0.0411
R_1	0.0338

5.2.40 [{PhC(N^{t}Bu)$_2$}{(*S*)-PEBA}Si] (**38**)

Compound	**38**
Formula	$C_{38}H_{46}N_4Si$
$D_{calc.}$/ g cm^{-3}	1.147
μ/mm^{-1}	0.101
Formula Weight	586.88
Colour	colourless
Shape	block
T/K	100
Crystal System	monoclinic
Flack Parameter	0.10(7)
Hooft Parameter	0.07(5)
Space Group	$P2_1$
a/Å	10.600(2)
b/Å	12.530(3)
c/Å	12.810(3)
α/°	
β/°	92.80(3)
γ/°	
V/Å^3	1699.4(6)
Z	2
Z'	1
Wavelength/Å	0.71073
Radiation type	MoK$_\alpha$
Θ_{min}/°	1.592
Θ_{max}/°	31.424
Measured Refl.	18515
Independent Refl.	8801
Reflections with I > 2(I)	6936
R_{int}	0.0329
Largest Peak	0.280
Deepest Hole	-0.197
GooF	1.018
wR_2 (all data)	0.1019
wR_2	0.0936
R_1 (all data)	0.0672
R_1	0.0452

5.2.41 [{(*S*)-PEBA}GeCl] (**39**)

Compound	**39**
Formula	$C_{23}H_{23}ClGeN_2$
$D_{calc.}$/ g cm^{-3}	1.367
μ/mm^{-1}	1.584
Formula Weight	435.47
Colour	colourless
Shape	prism
T/K	100
Crystal System	orthorhombic
Flack Parameter	-0.021(8)
Hooft Parameter	-0.006(4)
Space Group	$P2_12_12_1$
a/Å	8.240(2)
b/Å	15.390(3)
c/Å	16.680(3)
α/°	
β/°	
γ/°	
V/Å^3	2115.3(7)
Z	4
Z'	1
Wavelength/Å	0.71073
Radiation type	Mo K_α
Θ_{min}/°	1.800
Θ_{max}/°	31.625
Measured Refl.	10676
Independent Refl.	5674
Reflections with I > 2(I)	5079
R_{int}	0.0263
Largest Peak	0.824
Deepest Hole	-0.290
GooF	0.850
wR_2 (all data)	0.1088
wR_2	0.1020
R_1 (all data)	0.0420
R_1	0.0355

6 Summary (Zusammenfassung)

6.1 Summary

In this thesis, chiral iminophosphonamine ligands (Figure 6.1) and corresponding metal complexes are reported. The first part of the thesis deals with the synthesis of new chiral iminophosphonamines with varying steric bulk and chiral side groups. Subsequent treatment of these chiral ligands with alkali metal precursors leads to the formation of their corresponding alkali metal derivatives (Schemes 6.1-6.3).

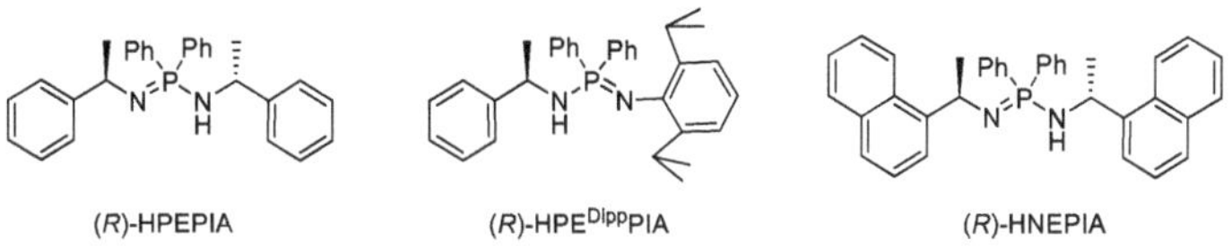

(*R*)-HPEPIA (*R*)-HPE^{Dipp}PIA (*R*)-HNEPIA

Figure 6.1 The chiral iminophosphonamines (*R*)-HPEPIA, (*R*)-HPE^{Dipp}PIA and (*R*)-HNEPIA.

More precisely, treatment of (*R*)-HPEPIA with KH, Rb or $RbN(SiMe_3)_2$ and Cs or $CsN(SiMe_3)_2$ resulted in the formation of the corresponding alkali metal complexes (**1**-**3**) (Scheme 6.1).

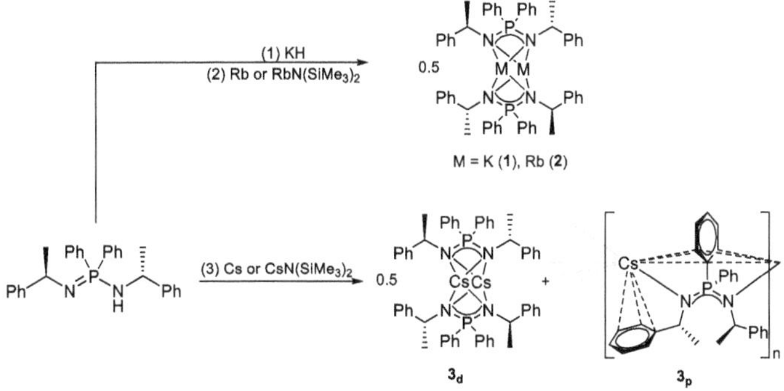

Scheme 6.1 Synthesis of alkali metal complexes (**1-3**) of (*R*)-HPEPIA.

The solid state structure determination by single crystal X-ray diffraction studies showed that all of these complexes adopts dimeric forms. Interestingly, due to the larger ionic radius of the caesium ion, the corresponding complex co-crystallized in both dimeric ($\mathbf{3_d}$) as well as polymeric ($\mathbf{3_p}$) forms.

By investigation of the photoluminescence (PL) studies of (*R*)-HPEPIA and the corresponding alkali metal complexes $[\{(R)\text{-PEPIA}\}M]_2$ (M = K (**1**), Rb (**2**) and Cs (**3**)) revealed quantum yields ranging from 8 to 36% at room temperature. Furthermore, temperature dependent PL measurements revealed that the (*R*)-HPEPIA ligand solely shows fluorescence under cryogenic conditions. For the alkali metal complexes a remarkably long-lived phosphorescence at low temperatures and an emission with a decay time of several microseconds was observed at room temperature. Additionally, all complexes show a non-uniform decrease of the PL intensity. These observations are characteristic for emission *via* a thermally activated delayed fluorescence (TADF) in the [{(*R*)-PEPIA}M]$_2$ complexes. The observed results were further supported by quantum chemical calculations. The study suggested that the interesting TADF behaviour of the complexes is arising due to their dimeric forms.

In order to study the coordination behaviour of (*R*)-HPE^{Dipp}PIA with alkali metals, the former was treated with [M{N(SiMe$_3$)$_2$}] (M = Li, Na), which resulted in the formation of the dimeric complexes, [{(*R*)-PE^{Dipp}PIA}$_2$M$_2$] (M = Li (**4**), Na (**5**)) (Scheme 6.2). The corresponding X-ray structures showed that the central metals form three four-membered rings fused in a stair-shaped structure involving the {(*R*)-PE^{Dipp}PIA}$^-$ ligand. The flanking Dipp groups adopt a *trans*-configuration with respect to the central metals.

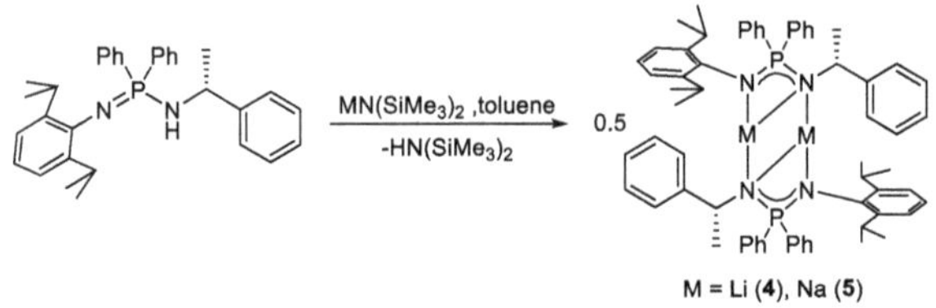

Ph Ph N=P N H → MN(SiMe$_3$)$_2$,toluene / -HN(SiMe$_3$)$_2$ → 0.5 [dimer: Ph Ph P N N M M N N P Ph Ph] M = Li (**4**), Na (**5**)

Scheme 6.2 Synthesis of [{(*R*)-PE^{Dipp}PIA}$_2$Li$_2$] (**4**) and [{(*R*)-PE^{Dipp}PIA}$_2$Na$_2$] (**5**).

Furthermore, the structural variation of the nitrogen substituents and the study of the coordination behaviour with alkali metals were extended to another ligand, (*R*)-HNEPIA. Treatment of (*R*)-HNEPIA with the corresponding alkali metal precursors yielded the monomeric complex [{(*R*)-NEPIA}Li(thf)$_2$] (**6**), and the dimeric complexes [{(*R*)-NEPIA}$_2$Na$_2$] (**7**) and [{(*R*)-NEPIA}$_2$K$_2$] (**8**) (Scheme 6.3). The molecular structures in the solid state showed that, in contrast to the monomeric complex **6**, the dimeric complexes **7** and **8** exhibit M···C interactions with the naphthyl substituents, similarly to the complexes **1-3**.

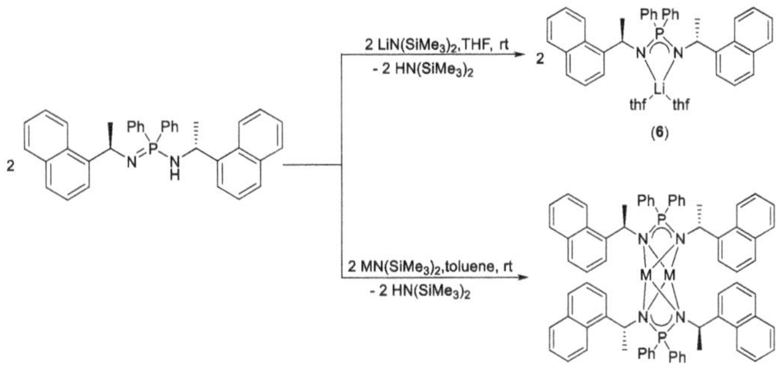

Scheme 6.3 Synthesis of [{(*R*)-NEPIA}Li(thf)$_2$] (**6**) and [{(*R*)-NEPIA}$_2$M$_2$] (M = Na (**7**), K (**8**)).

Thus, the observations by X-ray structure analysis indicate that the aggregation of the alkali metal complexes depends upon both, the ionic radius of the central metal as well as the steric demand of the substituents.

The interesting coordination and photophysical properties of the iminophosphonamides with alkali metals attracted analogous investigations with alkaline earth metals and divalent ytterbium complexes. In this regard, the complexes [{(*R*)-PEPIA}$_2$Ca] (**9**), [{(*R*)-PE^{Dipp}PIA}{N(SiMe$_3$)$_2$}Mg(thf)] (**10**), [{(*R*)-PE^{Dipp}PIA}$_2$Mg] (**11**), [{(*R*)-PE^{Dipp}PIA}$_2$Ca] (**12**), [{(*R*)-PE^{Dipp}PIA}$_2$Yb] (**13**), and [{(*R*)-NEPIA}$_2$Ca] (**14**) were synthesized by base elimination of (*R*)-HPEPIA, (*R*)-HPE^{Dipp}PIA and (*R*)-HNEPIA with [M{N(SiMe$_3$)$_2$}(thf)$_2$] (M = Mg, Ca, Yb) (Scheme 6.4). X-ray structure analysis of complexes **9-14** showed that the central metals adopt distorted tetrahedral geometries in all cases. In order to minimise steric repulsion with the ligand substituents, the central metals slightly deviate from the plane defined by the NPN ligand framework. The activity of the tetracoordinated homoleptic complexes **9**, **12** and **14** as catalysts in intermolecular hydroboration of ketones was further investigated. Rapid conversion of ketones into the corresponding alkoxyboronate esters was observed in quantitative yield. The high catalytic activity is possibly due to the coordinatively unsaturated nature of these complexes. Furthermore, the photoluminescence properties of the calcium complexes **9**, **12** and **14** were investigated. Whereby similar temperature dependent behaviour as observed for the alkali metal complexes was characterized.

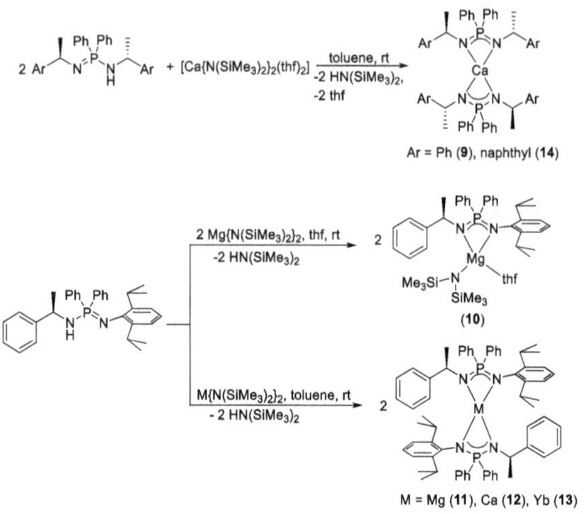

Scheme 6.4 Syntheses of complexes **9-14**.

The PL spectra of **9** and **12** show contributions of msec-long phosphorescence and thermally activated delayed fluorescence (TADF), with a strong variation of the PL lifetimes over temperatures ranging from 5 to 295 K. In contrast, the emission of **14** is assigned to fluorescence, which is likely arising from the naphthyl groups.

In order to further deepen the understanding of the PL properties of the iminophosphonamide ligand framework, transition metal complexes were synthesized. Hereby focus was put on d^{10} metal complexes since metal cores with fully filled d-orbitals are known to show interesting PL properties. In this regard, a variety of substituted iminophosphonamide ligands has been selected and their complexation behaviour towards Cu(I) and Zn(II) ions was studied (Figure 6.2). The solid-state structures of the copper complexes **15**, **17** and **18** showed intermetallic Cu···Cu distances lying within the range of cuprophilic interactions. The corresponding zinc complexes **16** and **19** showed distortion from their respective ideal tetrahedral and trigonal planar geometries due to the chelating nature of the iminophosphonamides. Temperature dependent photoluminescence studies revealed that the copper complexes **15** and **18** show TADF at temperatures above 140 K, while complex **17** displays a combination of fluorescence and phosphorescence on particular wavelengths. In contrast, the zinc complexes **16** and **19** merely show fluorescence at room temperature.

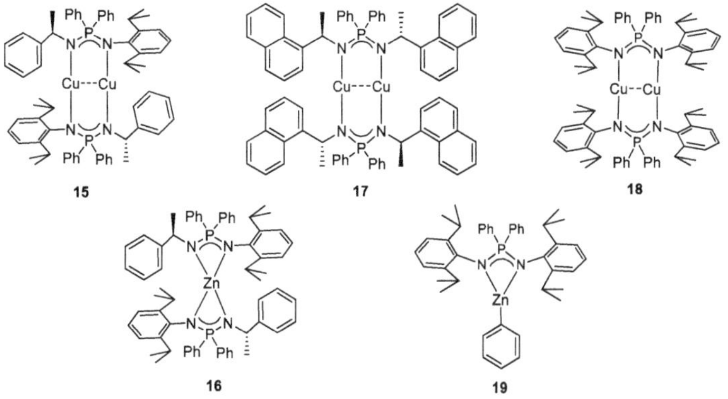

Figure 6.2 Copper and zinc complexes (**15-19**) of substituted iminophosphonamides.

In further coordination studies, *i.e.* the synthesis and structural investigation of group 13 (Al and Ga) complexes bearing chiral iminophosphonamides was carried out. In this regard, the symmetrical (*R*)-HPEPIA and unsymmetrical (*R*)-HPE^{Dipp}PIA ligand precursors were selected. Complexes **20-22** based on the bulkier [(*R*)-PE^{Dipp}PIA]$^-$ ligand yielded stable monosubstituted products, whereas the complexes **23-27** based on the [(*R*)-PEPIA]$^-$ ligand afforded disubstituted products (Schemes 6.5).

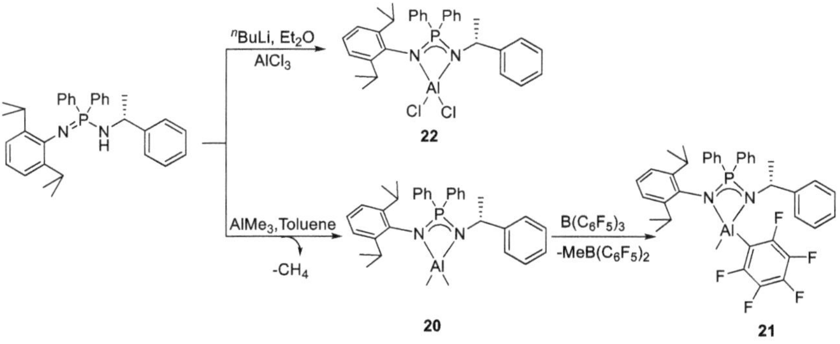

Scheme 6.5 Synthesis of complexes **20-22**.

In addition, the attempted synthesis of a complex with a weakly coordinating cation using [{(*R*)-PE^{Dipp}PIA}AlMe$_2$] (**20**) resulted in the formation of the unexpected neutral complex [{(*R*)-PE^{Dipp}PIA}AlMe(C_6F_5)] (**21**). Nevertheless, the halide abstraction method on [{(*R*)-PEPIA}$_2$MCl] (M = Al (**23**), Ga (**24**)) afforded the expected cationic complexes (**25** and **26**) (Figure 6.3).

These results suggest that the formation of mono- and disubstituted complexes highly depends on the steric influence of the nitrogen substituents at the iminophosphonamide ligand. Unfortunately, preliminary test reactions of these complexes in Lewis acid mediated reactions did not result in a noticeable catalytic activity. Such a lack of activity may be due to partial quenching of the Lewis acidity by the strong donating nature of the NPN ligands.

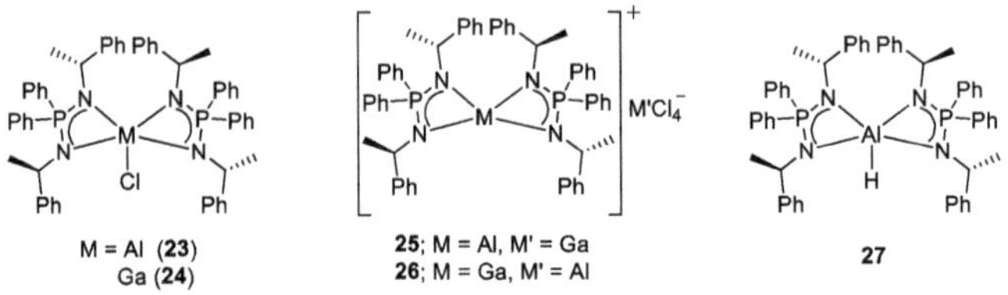

M = Al (**23**)
Ga (**24**)

25; M = Al, M' = Ga
26; M = Ga, M' = Al

27

Figure 6.3 Structures of complexes **23-27**.

Furthermore, using the chiral ligands discussed above, enantiopure group 4 metal complexes (**28-30**) were synthesized and structurally characterized (Figure 6.4). Owing to the similar ionic radii of Zr and Hf, complexes **29** and **30** are isostructural.

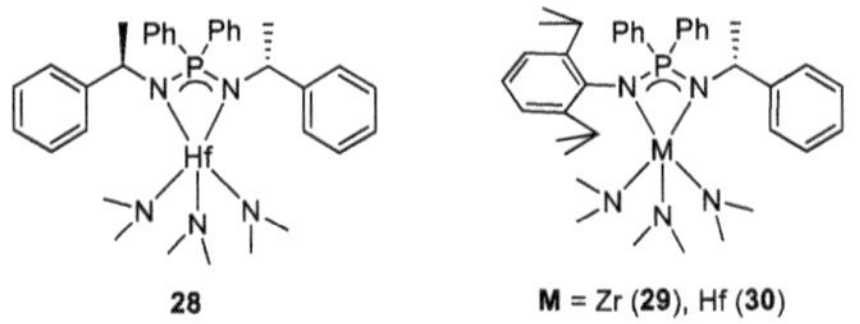

28

M = Zr (**29**), Hf (**30**)

Figure 6.4 Structures of complexes **28-30** obtained *via* base elimination.

Moreover, the synthesis and characterization of homoleptic lanthanide complexes of (*R*)-HPEPIA was investigated. Accordingly, the salt metathesis reaction between $LnCl_3$ and [{(*R*)-PEPIA}$_2$K$_2$] (**1**) resulted in the formation of isostructural homoleptic complexes [{(*R*)-PEPIA}$_3$Ln] (Ln = Y (**31**), La (**32**), Tb (**33**), and Yb (**34**) and Lu (**35**)) (Scheme 6.6).

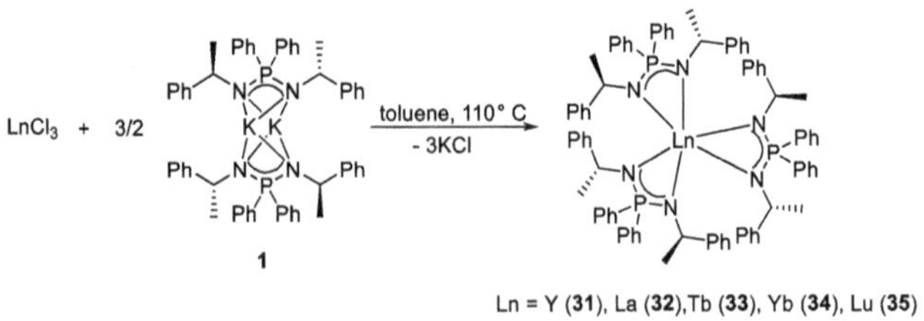

$LnCl_3$ + 3/2 **1** → toluene, 110° C, - 3KCl

Ln = Y (**31**), La (**32**),Tb (**33**), Yb (**34**), Lu (**35**)

Scheme 6.6 Synthesis of homoleptic lanthanide complexes (**31-35**).

All of these complexes, crystallize in form of propeller-type structures. Finally approaches to investigate the reactivity of iminophosphonamide and amidinate ligands towards tetrylenes were carried out. The reaction of [{(*R*)-PE^{Dipp}PIA}$_2$Li$_2$] (**4**) with [{PhC(^{t}BuN)$_2$}SiCl] in an equimolar ratio resulted into the isolation of an unexpected product, [{PhC(N^{t}Bu)$_2$}Si(=NDipp){N(*R*)(CH)MePh(PPh$_2$)}] (**36**). Formation of **36** is attributed to P-N bond activation of **4** by the reactive [{PhC(^{t}BuN)$_2$}SiCl] and possibly occurs through oxidative addition on an amidosilylene intermediate. To further investigate the reactivity with heavier tetrylenes, the reaction between $GeCl_2$·dioxane and **4** was performed in an equimolar ratio and yielded the expected transmetallation product **37** (Scheme 6.7).

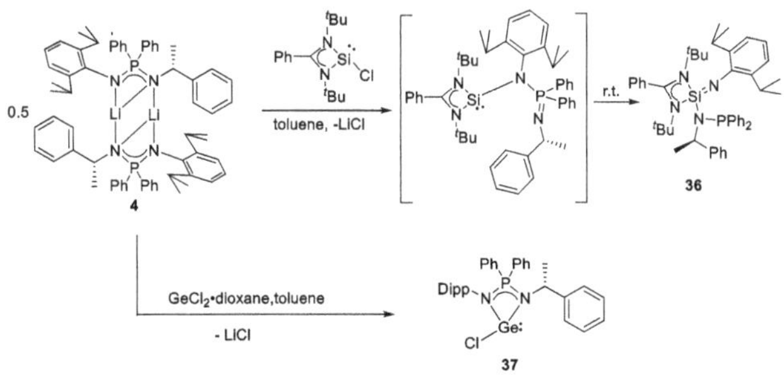

Scheme 6.7 Reactivity of tetrylenes with the iminophosphonamide **4**.

In contrast to the unexpected formation of **36**, the reaction of [{PhC(^{t}BuN)$_2$}SiCl] with the amidinate (*S*)-LiPEBA resulted into the formation of the expected product **38** (Scheme 6.8). This result highlights the lower reactivity of (*S*)-LiPEBA as compared to that of **4**, the latter being prone to P-N bond activation reactions due to the presence of a reducible P(V) centre. Furthermore, the salt metathesis reaction between $GeCl_2$·dioxane and (*S*)-LiPEBA yielded the monochlorogermylene **39**, which crystallizes structurally similar to the iminophosphonamide germylene **37**.

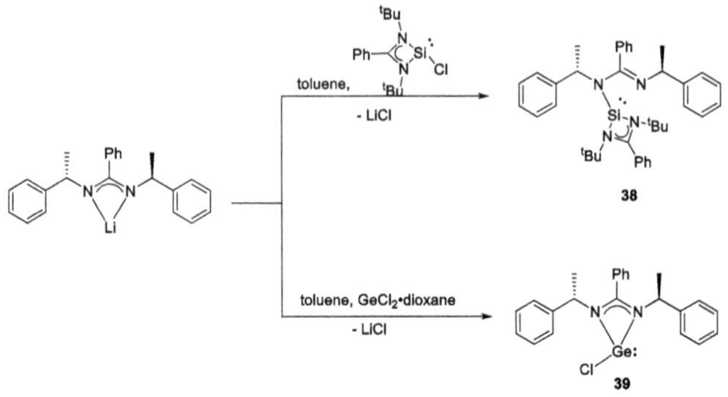

Scheme 6.8 Reactivity of tetrylenes with the compound (*S*)-LiPEBA.

6.2 Zusammenfassung

Die vorliegende Arbeit behandelt die Synthese sowie Charakterisierung von Metallkomplexen dreier verschiedener chiraler Iminophosphonamine (Abbildung 6.1). Der erste Teil der Arbeit beschreibt hierbei die Synthese der neuen chiralen Iminophosphonamine, welche im sterischen Anspruch und den chiralen Resten variieren. Anschließend wurden die Umsetzungen der erhaltenen Liganden zu den entsprechenden Alkalimetallkomplexen untersucht (Schema 6.1-6.3).

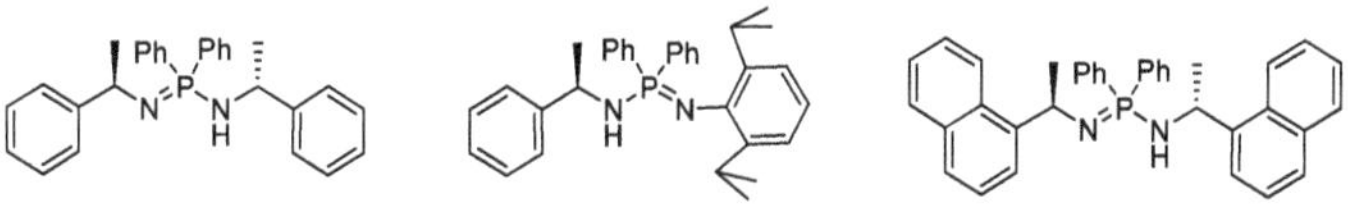

Abbildung 6.1 Chirale Iminophosphonamine (*R*)-HPEPIA, (*R*)-HPE^{Dipp}PIA und (*R*)-HNEPIA.

Die Reaktion von (*R*)-HPEPIA mit KH, Rb oder $RbN(SiMe_3)_2$ sowie Cs oder $CsN(SiMe_3)_2$ führte zu den Verbindungen **1**-**3** (Schema 6.1).

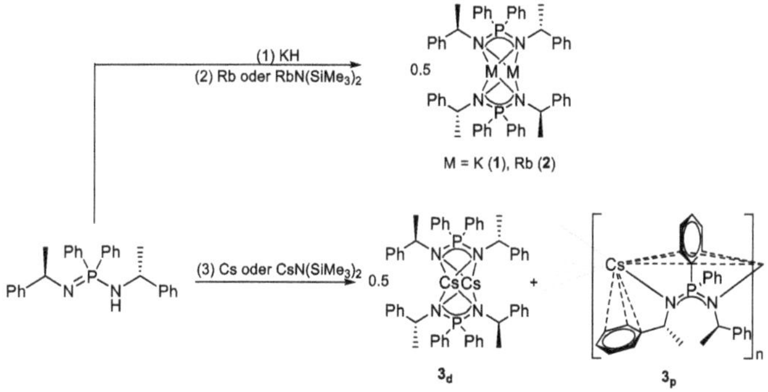

Schema 6.1 Synthese der Alkalimetallkomplexe (**1**-**3**) von (*R*)-HPEPIA.

Während die Kalium- und Rubidium-Verbindungen (**1**, **2**) im Festkörper als Dimere vorliegen, co-kristallisierte für Verbindung **3**, aufgrund des größeren Ionenradiuses des Cäsiums, neben der dimeren Struktur ($\mathbf{3_d}$) auch eine polymere Form ($\mathbf{3_p}$).

Photolumineszenz-Untersuchungen (PL) von (*R*)-HPEPIA und den erhaltenen Komplexen [{(*R*)-PEPIA}M]$_2$ (M = K (**1**), Rb (**2**) und Cs (**3**)) zeigten bei Raumtemperatur eine Quantenausbeute von 8 bis 36 %. Temperaturabhängige PL-Messungen des (*R*)-HPEPIA Liganden zeigten zudem eine Fluoreszenz unter kryogenen Temperaturen. Für die Alkalimetallkomplexe konnte hingegen eine langlebige Phosphoreszenz bei tiefen Temperaturen und eine Emission mit einer Abklingzeit von mehreren Mikrosekunden bei Raumtemperatur beobachtet werden. Darüber hinaus zeigten alle Komplexe eine ungleichmäßige Abnahme der PL Intensität. Diese Beobachtungen in den PL-Untersuchungen der [{(*R*)-PEPIA}M]$_2$ Komplexe sind charakteristisch für thermisch aktivierte verzögerte Fluoreszenz (TADF). Weiterführende quantenmechanische Rechnungen hierzu lassen darauf schließen, dass dieses interessante TADF-Verhalten der Komplexe auf ihre dimere Form zurückzuführen ist.

Im weiteren Verlauf der Arbeit wurde das Koordinationsverhalten von (*R*)-HPE^{Dipp}PIA und (*R*)-HNEPIA gegenüber den Alkalimetallen untersucht. Die Umsetzung von (*R*)-HPE^{Dipp}PIA mit [M{N(SiMe$_3$)$_2$}] (M = Li, Na), führte hierbei zu den dimeren Verbindungen [{(*R*)-PE^{Dipp}PIA}$_2$M$_2$] (M = Li (**4**), Na (**5**)) (Schema 6.2). Im Festkörper bilden die zentralen Metallatome unter Einbeziehung der zwei {(*R*)-PE^{Dipp}PIA}$^-$ Liganden eine stufenförmige Struktur aus, welche aus drei viergliedrigen Ringen zusammengesetzt ist. Die Dipp-Substituenten nehmen hierbei gegenüber den zentralen Metallen eine trans-Konfiguration ein.

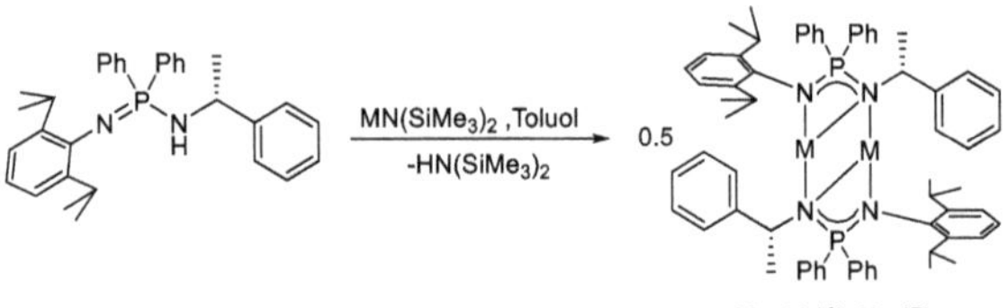

Schema 6.2 Synthese von [{(*R*)-PE^{Dipp}PIA}$_2$Li$_2$] (**4**) und [{(*R*)-PE^{Dipp}PIA}$_2$Na$_2$] (**5**).

Die Umsetzung von (*R*)-HNEPIA mit den entsprechenden Alkalimetallvorläuferverbindungen führte zu dem monomeren Komplex [{(*R*)-NEPIA}Li(thf)$_2$] (**6**) sowie den dimeren Strukturen [{(*R*)-NEPIA}$_2$Na$_2$] (**7**) und [{(*R*)-NEPIA}$_2$K$_2$] (**8**) (Schema 6.3). Hierbei weisen die dimeren Verbindungen **7** und **8** im Festkörper eine M···C Koordination durch die Naphtyl-Substituenten auf (analog zu den Verbindungen **1**-**3**).

Insgesamt konnte somit eine Abhängigkeit der Aggregation der Alkalimetallkomplexe sowohl vom ionischen Radius des verwendeten Metalls als auch vom sterischen Anspruch des jeweiligen Liganden festgestellt werden.

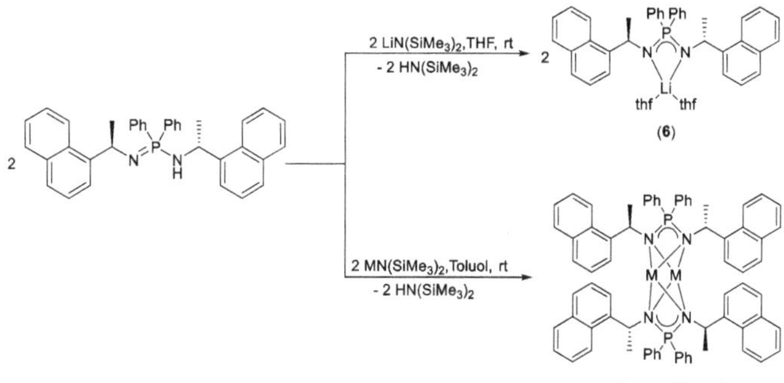

Schema 6.3 Synthese von [{(*R*)-NEPIA}Li(thf)$_2$] (**6**) und [{(*R*)-NEPIA}$_2$M$_2$] (M = Na (**7**), K (**8**)).

Aufgrund dieser Ergebnisse und den interessanten photophysikalischen Eigenschaften wurden im weiteren Verlauf der Arbeit Synthese und Eigenschaften der entsprechenden Erdalkalimetallkomplexe sowie eines divalenten Ytterbiumkomplexes untersucht. Hierbei wurden die Verbindungen [{(*R*)-PEPIA}$_2$Ca] (**9**), [{(*R*)-PE^{Dipp}PIA}{N(SiMe$_3$)$_2$}Mg(thf)] (**10**), [{(*R*)-PE^{Dipp}PIA}$_2$Mg] (**11**), [{(*R*)-PE^{Dipp}PIA}$_2$Ca] (**12**), [{(*R*)-PE^{Dipp}PIA}$_2$Yb] (**13**) und [{(*R*)-NEPIA}$_2$Ca] (**14**) über Baseneleminierungsreaktionen von (*R*)-HPEPIA, (*R*)-HPE^{Dipp}PIA und (*R*)-HNEPIA mit [M{N(SiMe$_3$)$_2$}(thf)$_2$] (M = Mg, Ca, Yb) erhalten (Schema 6.4).

Für alle erhaltenen Verbindungen (**9-14**) liegt im Festkörper eine verzerrt-tetraedrische Koordination des Metalls vor. Hierbei weicht das Metall leicht von der durch das NPN Ligandengerüst definierten Ebene ab, um die sterische Abstoßung mit den Ligandensubstituenten zu minimieren.

Die homoleptischen Komplexe **9, 12** und **14** wurden auf ihre Aktivität als Katalysatoren für die intermolekulare Hydroborierung von Ketonen untersucht. Hierbei konnte eine schnelle Umwandlung der Ketone zu den entsprechenden Alkoxy-Boronat-Estern in quantitativen Ausbeuten beobachtet werden. Diese hohe katalytische Aktivität lässt sich durch die fehlende koordinative Absättigung der Komplexe erklären. Weiterhin wurden die photophysikalischen Eigenschaften der Calciumkomplexe **9**, **12** und **14** untersucht. Hierbei konnte ein, zu den Alkalimetallkomplexen ähnliches, temperaturabhängiges Verhalten beobachtet werden. Die PL Spektren von **9** und **12** zeigen Beiträge einer Mikrosekunden langen Phosphoreszenz sowie TADF mit einer starken Variation der PL Lebenszeiten bei Temperaturen von 5 bis 295 K.

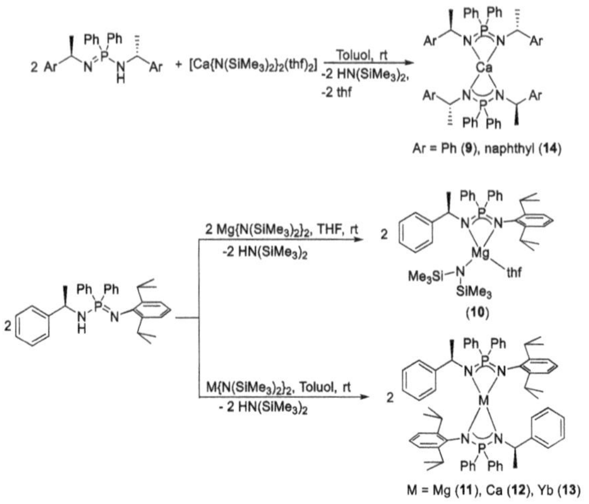

Schema 6.4 Synthese der Komplexe **9-14**.

Im Gegensatz hierzu konnte die Emission von **14** einer Fluoreszenz zugeordnet werden, welche sich vermutlich auf die Naphthyl-Gruppen zurückführen lässt.

Um das Verständnis der PL-Eigenschaften des Ligandengerüstes zu vertiefen, wurden entsprechende Übergangsmetallkomplexe synthetisiert. Der Fokus wurde hierbei auf d^{10}-Metallkomplexe gelegt, da solche Verbindungen mit vollständig gefüllten d-Orbitalen bekannt für ihre interessanten PL Eigenschaften sind. Die Umsetzung einer Auswahl verschieden substituierter Iminophosphonamide mit geeigneten Cu(I) und Zn(II) Vorläuferverbindungen ergab die Komplexe **15-19** (Abbildung 6.2).

Die Festkörperstrukturen der Kupferkomplexe **15**, **17** und **18** zeigten intermetallische Cu···Cu Abstände im Bereich metallophiler Wechselwirkungen. Für die Zinkkomplexe **16** und **19** konnte im Festkörper aufgrund der chelatisierenden Liganden eine Verzerrung ihrer idealen tetraedrischen bzw. trigonal-planaren Geometrie festgestellt werden.

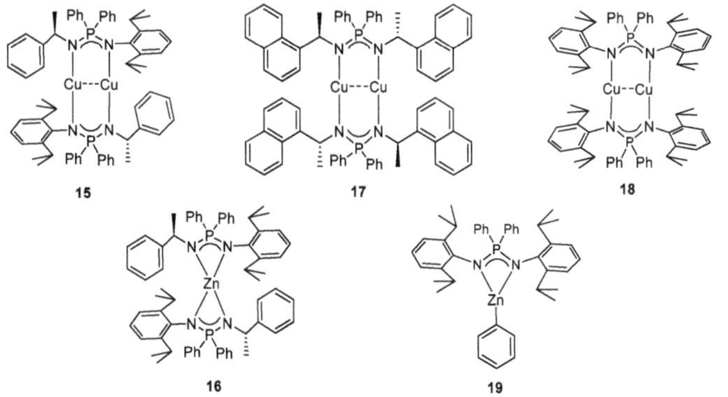

Abbildung 6.2 Synthetisierte Kupfer und Zink Komplexe (**15-19**) der Iiminophosphonamide.

Temperaturabhängige PL-Messungen zeigten ein TADF Verhalten für die Kupferkomplexe **15** und **18** bei Temperaturen über 140 K. Komplex **17** zeigte hingegen eine Kombination von Phosphoreszenz und Fluoreszenz für bestimmte Wellenlängen. Im Gegensatz hierzu zeigten die Zinkkomplexe **16** und **19** lediglich Fluoreszenz bei Raumtemperatur.

In weiteren Verlauf der Arbeit wurde die Synthese von chiralen Gruppe 13 (Al und Ga) Komplexen mit dem symmetrischen (*R*)-HPEPIA und dem unsymmetrischen (*R*)-HPE^{Dipp}PIA Liganden untersucht. Bei der Verwendung des sterisch anspruchsvolleren [(*R*)-PE^{Dipp}PIA]$^-$ Liganden wurden die monosubstituierten Produkte **20-22** erhalten, während für das kleinere [(*R*)-PEPIA]$^-$ die disubstituierten Verbindungen **23-27** erhalten wurden (Schema 6.5).

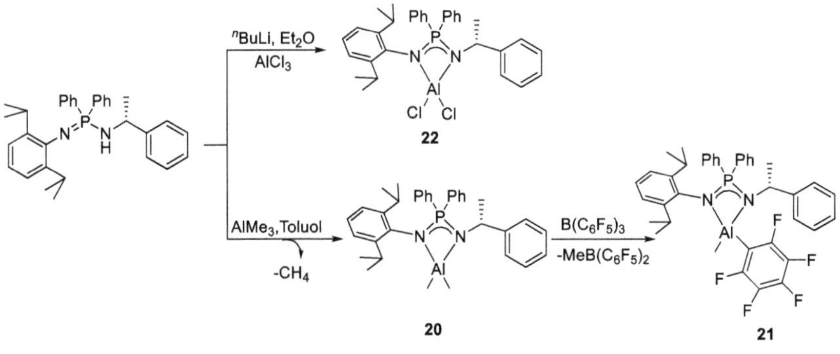

Schema 6.5 Synthese der Komplexe **20-22**.

Ergänzend wurde bei dem Versuch ausgehend von [{(*R*)-PE^{Dipp}PIA}AlMe$_2$] (**20**) einen Komplex mit einem schwach koordinierendem Kation darzustellen, der unerwartete, neutrale Komplex [{(*R*)-PE^{Dipp}PIA}AlMe(C_6F_5)] (**21**) erhalten. Des Weiteren führte die Halogenabstraktion an [{(*R*)-PEPIA}$_2$MCl] (M = Al (**23**), Ga (**24**)) zu den erwarteten kationischen Komplexen **25** und **26** (Abbildung 6.3). Testreaktionen mit diesen Verbindungen in Lewissäure-vermittelten Reaktionen ergaben keine signifikante katalytische Aktivität. Dies lässt sich unter Umständen durch die starken Donoreigenschaften des NPN-Ligadens erklären.

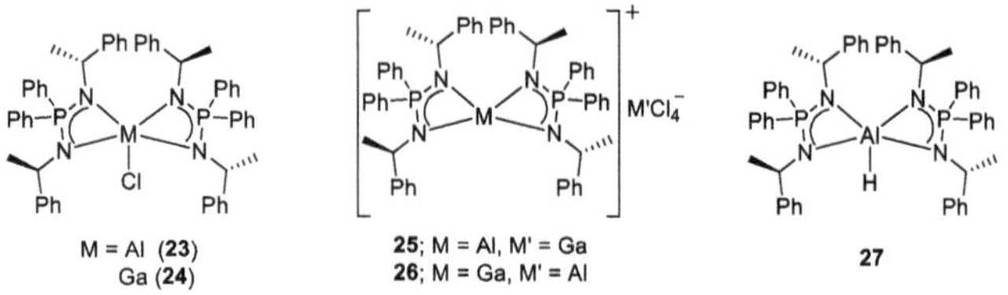

M = Al (**23**)
Ga (**24**)

25; M = Al, M' = Ga
26; M = Ga, M' = Al

27

Abbildung 6.3 Verbindungen **23-27**.

Darüber hinaus wurden die oben beschriebenen chiralen Liganden verwendet, um enantiomerenreine Komplexe der Gruppe 4 Metalle zu synthetisieren (**28-30**) (Abbildung 6.4). Hierbei sind die Verbindungen **29** und **30** aufgrund des ähnlichen Ionenradiuses von Zr und Hf isostrukturell.

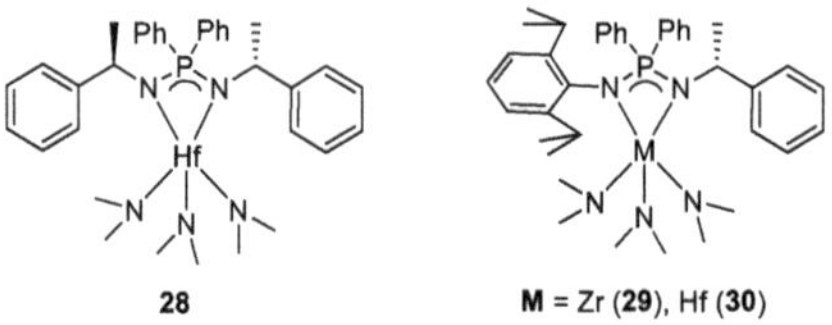

28

M = Zr (**29**), Hf (**30**)

Abbildung 6.4 Durch Baseneliminierung erhaltene Komplexe **28-30**.

Für die Synthese und Charakterisierung von homoleptischen Lanthanoidkomplexen wurde der Ligand (*R*)-HPEPIA gewählt. Die Salzmetathese von $LnCl_3$ mit [{(*R*)-PEPIA}$_2$K$_2$] (**1**) ergab eine Reihe isostruktureller Verbindungen [{(*R*)-PEPIA}$_3$Ln] (Ln = Y (**31**), La (**32**), Tb (**33**) und Yb (**34**), Lu (**35**)) (Schema 6.6). Im Festkörper liegen alle diese Komplexe in propellerartiger Struktur vor.

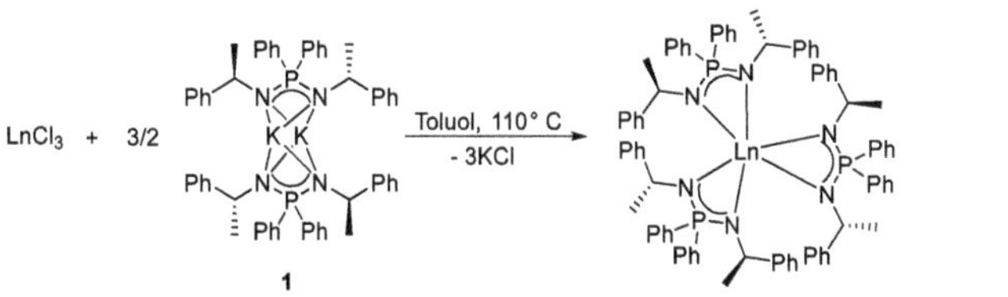

Ln = Y (**31**), La (**32**),Tb (**33**), Yb (**34**), Lu (**35**)

Schema 6.6 Synthese der homoleptischen Lanthanoidkomplexe (**31-35**).

Im letzten Teil der vorliegenden Arbeit wurde schließlich die Reaktivität von Iminophosphonamid- und Amidinat-Liganden gegenüber Tetrylenen untersucht. Die Reaktion von [{(*R*)-PE^{Dipp}PIA}$_2$Li$_2$] (**4**) mit [{PhC(^{t}BuN)$_2$}SiCl] in einem equimolaren Verhältnis ergab das unerwartete Produkt [{PhC(N^{t}Bu)$_2$}Si(=NDipp){N(*R*)(CH)MePh(PPh$_2$)}] (**36**). Dieses kann durch eine P-N Bindungsaktivierung von **4** mit dem reaktiven Komplex [{PhC(^{t}BuN)$_2$}SiCl] und oxidativer Addition am Amidosilylen-Intermediat erklärt werden. Im Gegensatz hierzu ergab die analoge Reaktion von **4** mit dem schwereren Tetrylen GeCl$_2$·dioxan das erwartete Transmetallierungsprodukt **37** (Schema 6.7).

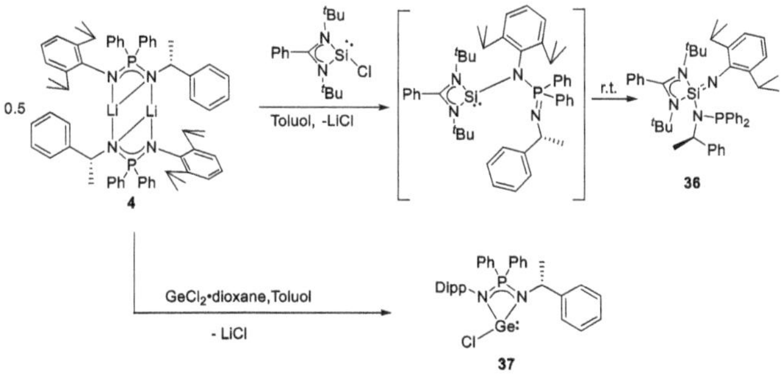

Schema 6.7 Reaktivität des Iminophosphonamides **4** gegenüber verschiedenen Tetrylenen.

Auch die Metathesereaktion von [{PhC(^{t}BuN)$_2$}SiCl] mit dem Amidinat (*S*)-LiPEBA resultierte in dem erwarteten Produkt **38** (Schema 6.8). Dies zeigt die geringere Reaktivität von (*S*)-LiPEBA im Vergleich zu **4**, in welchem die P-N Bindungsaktivierung durch das reduzierbare P(V) Zentrum ermöglicht wird. Abschließend ergab die Reaktion von GeCl$_2$·dioxan und (*S*)-LiPEBA das Monochlorogermylen **39**.

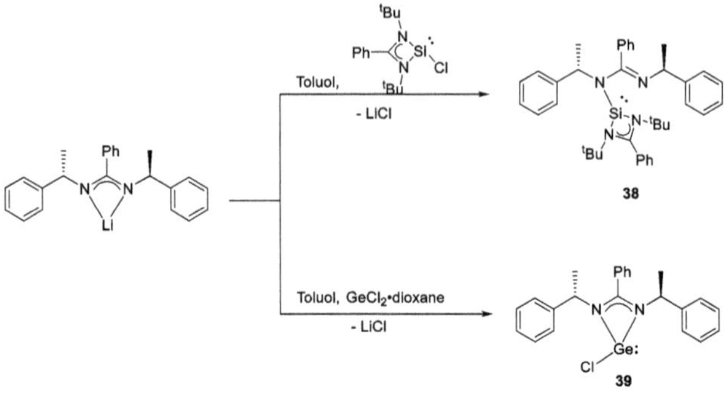
tBu
Ph
Si
Cl
Toluol,
- LiCl
Ph
N
Li
Toluol, GeCl2•dioxane
- LiCl
38
Ge:
Cl
39

Schema 6.8 Reaktivität von Tetrylenen mit (*S*)-LiPEBA.

7 References

1. (a) J. Barker and M. Kilner, *Coord. Chem. Rev.*, 1994, **133**, 219-300; (b) P. C. Junk and M. L. Cole, *Chem. Commun.*, 2007, 1579-1590.
2. S. S. Sen, S. Khan, S. Nagendran and H. W. Roesky, *Acc. Chem. Res.*, 2012, **45**, 578-587.
3. F. T. Edelmann, *Coord. Chem. Rev.*, 1994, **137**, 403-481.
4. (a) D. M. Grove, G. van Koten, H. J. C. Ubbels, K. Vrieze, L. C. Niemann and C. H. Stam, *J. Chem. Soc., Dalton Trans.*, 1986, 717-724; (b) F. A. Cotton, L. M. Daniels, D. J. Maloney, J. H. Matonic and C. A. Murillo, *Polyhedron*, 1994, **13**, 815-823; (c) J. A. R. Schmidt and J. Arnold, *J. Chem. Soc., Dalton Trans.*, 2002, 2890-2899.
5. S. R. Foley, C. Bensimon and D. S. Richeson, *J. Am. Chem. Soc.*, 1997, **119**, 10359-10363.
6. (a) M. J. R. Brandsma, E. A. C. Brussee, A. Meetsma, B. Hessen and J. H. Teuben, *Eur. J. Inorg. Chem.*, 1998, 1867-1870; (b) H. Kay Lee, T. Suet Lam, C.-K. Lam, H.-W. Li and S. Man Fung, *New J. Chem.*, 2003, **27**, 1310-1318; (c) J. Grundy, M. P. Coles and P. B. Hitchcock, *New J. Chem.*, 2004, **28**, 1195-1197.
7. A. R. Corcos and J. F. Berry, *Dalton Trans.*, 2017, **46**, 5532-5539.
8. (a) H. Shoukang, S. Gambarotta, C. Bensimon and J. J. H. Edema, *Inorg. Chim. Acta*, 1993, **213**, 65-74; (b) T. J. Feuerstein, M. Poß, T. P. Seifert, S. Bestgen, C. Feldmann and P. W. Roesky, *Chem. Commun.*, 2017, **53**, 9012-9015.
9. F. T. Edelmann, *Chem. Soc. Rev.*, 2012, **41**, 7657-7672.
10. F. A. Cotton, T. Inglis, M. Kilner and T. R. Webb, *Inorg. Chem.*, 1975, **14**, 2023-2026.
11. F. A. Cotton, L. M. Daniels, X. Feng, D. J. Maloney, J. H. Matonic and C. A. Murilio, *Inorg. Chim. Acta*, 1997, **256**, 291-301.
12. F. T. Edelmann, *Chem. Soc. Rev.*, 2009, **38**, 2253-2268.
13. (a) A. R. Sadique, M. J. Heeg and C. H. Winter, *Inorg. Chem.*, 2001, **40**, 6349-6355; (b) J. Barker, N. C. Blacker, P. R. Phillips, N. W. Alcock, W. Errington and M. G. H. Wallbridge, *J. Chem. Soc., Dalton Trans.*, 1996, 431-437; (c) D. Doyle, Y. K. Gun'ko, P. B. Hitchcock and M. F. Lappert, *J. Chem. Soc., Dalton Trans.*, 2000, 4093-4097.
14. M. P. Coles and R. F. Jordan, *J. Am. Chem. Soc.*, 1997, **119**, 8125-8126.
15. (a) S. S. Sen, A. Jana, H. W. Roesky and C. Schulzke, *Angew. Chem. Int. Ed.*, 2009, **48**, 8536-8538; (b) C.-W. So, H. W. Roesky, J. Magull and R. B. Oswald, *Angew. Chem. Int. Ed.*, 2006, **45**, 3948-3950; (c) S. Nagendran, S. S. Sen, H. W. Roesky, D. Koley, H. Grubmüller, A. Pal and R. Herbst-Irmer, *Organometallics*, 2008, **27**, 5459-5463; (d) S. S. Sen, M. P. Kritzler-Kosch, S. Nagendran, H. W. Roesky, T. Beck, A. Pal and R. Herbst-Irmer, *Eur. J. Inorg. Chem.*, 2010, 5304-5311.
16. X. Sun, T. Simler, R. Yadav, R. Köppe and P. W. Roesky, *J. Am. Chem. Soc.*, 2019, **141**, 14987-14990.
17. B. S. Lim, A. Rahtu, J.-S. Park and R. G. Gordon, *Inorg. Chem.*, 2003, **42**, 7951-7958.
18. M. Wedler, F. Knösel, U. Pieper, D. Stalke, F. T. Edelmann and H.-D. Amberger, *Chem. Ber.*, 1992, **125**, 2171-2181.
19. Y. Luo, Y. Yao, Q. Shen, J. Sun and L. Weng, *J. Organomet. Chem.*, 2002, **662**, 144-149.
20. R. Duchateau, C. T. van Wee, A. Meetsma and J. H. Teuben, *J. Am. Chem. Soc.*, 1993, **115**, 4931-4932.
21. (a) M. Wedler, H. W. Roesky and F. Edelmann, *J. Organomet. Chem.*, 1988, **345**, C1-C3; (b) M. Wedler, F. Knösel, M. Noltemeyer, F. T. Edelmann and U. Behrens, *J. Organomet. Chem.*, 1990, **388**, 21-45; (c) I. S. R. Karmel, N. Fridman and M. S. Eisen, *Organometallics*, 2015, **34**, 636-643.
22. (a) C. Fedorchuk, M. Copsey and T. Chivers, *Coord. Chem. Rev.*, 2007, **251**, 897-924; (b) S. Harder, *Dalton Trans.*, 2010, **39**, 6677-6681.
23. (a) S.-O. Hauber, F. Lissner, G. B. Deacon and M. Niemeyer, *Angew. Chem. Int. Ed.*, 2005, **44**, 5871-5875; (b) N. Nimitsiriwat, V. C. Gibson, E. L. Marshall, P. Takolpuckdee, A. K. Tomov, A.

J. P. White, D. J. Williams, M. R. J. Elsegood and S. H. Dale, *Inorg. Chem.*, 2007, **46**, 9988-9997; (c) S.-O. Hauber and M. Niemeyer, *Inorg. Chem.*, 2005, **44**, 8644-8646; (d) S. G. Alexander, M. L. Cole, C. M. Forsyth, S. K. Furfari and K. Konstas, *Dalton Trans.*, 2009, 2326-2336; (e) C. GuhaRoy, R. J. Butcher and S. Bhattacharya, *J. Organomet. Chem.*, 2008, **693**, 3923-3931.

24. (a) M. M. Meinholz, S. K. Pandey, S. M. Deuerlein and D. Stalke, *Dalton Trans.*, 2011, **40**, 1662-1671; (b) F. Pauer, J. Rocha and D. Stalke, *J. Chem. Soc., Chem. Commun.*, 1991, 1477-1479; (c) R. Fleischer, B. Walfort, A. Gbureck, P. Scholz, W. Kiefer and D. Stalke, *Chem. Eur. J.*, 1998, **4**, 2266-2274; (d) F. T. Edelmann, F. Knösel, F. Pauer, D. Stalke and W. Bauer, *J. Organomet. Chem.*, 1992, **438**, 1-10.
25. (a) A. Steiner, S. Zacchini and P. I. Richards, *Coord. Chem. Rev.*, 2002, **227**, 193-216; (b) M. Witt and H. W. Roesky, *Chem. Rev.*, 1994, **94**, 1163-1181; (c) B. Prashanth and S. Singh, *Dalton Trans.*, 2014, **43**, 16880-16888.
26. A. Stasch, *Chem. Eur. J.*, 2012, **18**, 15105-15112.
27. C. M. D. Komen, *Recl. Trav. Chim. Pays-Bas*, 1994, **113**, 382-382.
28. H. Staudinger and J. Meyer, *Helv. Chim. Acta*, 1919, **2**, 635-646.
29. A. L. Hawley and A. Stasch, *Eur. J. Inorg. Chem.*, 2015, 258-270.
30. B. Prashanth and S. Singh, *J. Chem. Sci.*, 2015, **127**, 315-325.
31. S. A. Ahmed, M. S. Hill, P. B. Hitchcock, S. M. Mansell and O. St John, *Organometallics*, 2007, **26**, 538-549.
32. A. Steiner and D. Stalke, *Inorg. Chem.*, 1993, **32**, 1977-1981.
33. R. Fleischer and D. Stalke, *Inorg. Chem.*, 1997, **36**, 2413-2419.
34. A. Stasch, *Angew. Chem. Int. Ed.*, 2014, **53**, 10200-10203.
35. S. R. Lawrence, D. B. Cordes, A. M. Z. Slawin and A. Stasch, *Dalton Trans.*, 2019, **48**, 16936-16942.
36. S. Wingerter, M. Pfeiffer, A. Murso, C. Lustig, T. Stey, V. Chandrasekhar and D. Stalke, *J. Am. Chem. Soc.*, 2001, **123**, 1381-1388.
37. S. Dagorne and S. Woodward, *Modern Organoaluminum Reagents: Preparation, Structure, Reactivity and Use*, Springer-Verlag Berlin Heidelberg, 2013.
38. (a) Z. Yang, M. Zhong, X. Ma, K. Nijesh, S. De, P. Parameswaran and H. W. Roesky, *J. Am. Chem. Soc.*, 2016, **138**, 2548-2551; (b) Z. Yang, M. Zhong, X. Ma, S. De, C. Anusha, P. Parameswaran and H. W. Roesky, *Angew. Chem. Int. Ed.*, 2015, **54**, 10225-10229.
39. A. Ford and S. Woodward, *Angew. Chem. Int. Ed.*, 1999, **38**, 335-336.
40. Y. Hayashi, J. J. Rohde and E. J. Corey, *J. Am. Chem. Soc.*, 1996, **118**, 5502-5503.
41. L. A. Berben, *Chem. Eur. J.*, 2015, **21**, 2734-2742.
42. (a) B. Nekoueishahraki, H. W. Roesky, G. Schwab, D. Stern and D. Stalke, *Inorg. Chem.*, 2009, **48**, 9174-9179; (b) H. Schmidbaur, K. Schwirten and H.-H. Pickel, *Chem. Ber.*, 1969, **102**, 564-567.
43. H. Allcock, *Phosphorus Nitrogen Compounds*, Academic Press, New York, 1972.
44. B. Prashanth, N. K. Srungavruksham and S. Singh, *ChemistrySelect*, 2016, **1**, 3601-3606.
45. J. Vrána, R. Jambor, A. Růžička, M. Alonso, F. De Proft and L. Dostál, *Eur. J. Inorg. Chem.*, 2014, 5193-5203.
46. A. L. Hawley, C. A. Ohlin, L. Fohlmeister and A. Stasch, *Chem. Eur. J.*, 2017, **23**, 447-455.
47. D. H. Harris and M. F. Lappert, *J. Chem. Soc., Chem. Commun.*, 1974, 895-896.
48. (a) S. S. Sen, H. W. Roesky, D. Stern, J. Henn and D. Stalke, *J. Am. Chem. Soc.*, 2010, **132**, 1123-1126; (b) P. Benndorf, C. Preuß and P. W. Roesky, *J. Organomet. Chem.*, 2011, **696**, 1150-1155.
49. (a) B. Prashanth and S. Singh, *Dalton Trans.*, 2016, **45**, 6079-6087; (b) O. A. Brusylovets, O. V. Vinichenko, A. I. Brusilovets, T. Lis, E. Bonnefille, S. Mazières and C. Couret, *Polyhedron*, 2010, **29**, 3269-3276; (c) U. Kilimann, M. Noltemeyer and F. T. Edelmann, *J. Organomet. Chem.*, 1993, **443**, 33-42.

50. (a) C.-Y. Qi and Z.-X. Wang, *J. Polym. Sci. Pole. Chem.*, 2006, **44**, 4621-4631; (b) J. Vrána, S. Ketkov, R. Jambor, A. Růžička, A. Lyčka and L. Dostál, *Dalton Trans.*, 2016, **45**, 10343-10354.
51. H. Nagashima, H. Kondo, T. Hayashida, Y. Yamaguchi, M. Gondo, S. Masuda, K. Miyazaki, K. Matsubara and K. Kirchner, *Coord. Chem. Rev.*, 2003, **245**, 177-190.
52. T. y. A. Peganova, O. A. Filippov, N. V. Belkova, I. V. Fedyanin and A. M. Kalsin, *Eur. J. Inorg. Chem.*, 2018, 5098-5107.
53. (a) W. Keim, R. Appel, A. Storeck, C. Krüger and R. Goddard, *Angew. Chem. Int. Ed.*, 1981, **20**, 116-117; (b) A. Dworak, W. Walach and B. Trzebicka, *Macromol. Chem. Phys.*, 1995, **196**, 1963-1970.
54. (a) B. F. Straub, F. Rominger and P. Hofmann, *Organometallics*, 2000, **19**, 4305-4309; (b) B. F. Straub and P. Hofmann, *Angew. Chem. Int. Ed.*, 2001, **40**, 1288-1290.
55. K. Albahily, S. Licciulli, S. Gambarotta, I. Korobkov, R. Chevalier, K. Schuhen and R. Duchateau, *Organometallics*, 2011, **30**, 3346-3352.
56. R. L. Stapleton, J. Chai, N. J. Taylor and S. Collins, *Organometallics*, 2006, **25**, 2514-2524.
57. B. F. Straub, F. Eisenträger and P. Hofmann, *Chem. Commun.*, 1999, 2507-2508.
58. H. Ackermann, O. Bock, U. Müller and K. Dehnicke, *Z. Anorg. Allg. Chem.*, 2000, **626**, 1854-1856.
59. H. L. Hermann, G. Boche and P. Schwerdtfeger, *Chem. Eur. J.*, 2001, **7**, 5333-5342.
60. F. Baier, Z. Fei, H. Gornitzka, A. Murso, S. Neufeld, M. Pfeiffer, I. Rüdenauer, A. Steiner, T. Stey and D. Stalke, *J. Organomet. Chem.*, 2002, **661**, 111-127.
61. T. y. A. Peganova, I. S. Sinopalnikova, A. S. Peregudov, I. V. Fedyanin, A. Demonceau, N. A. Ustynyuk and A. M. Kalsin, *Dalton Trans.*, 2016, **45**, 17030-17041.
62. D. Fenske, B. Maczek and K. Maczek, *Z. Anorg. Allg. Chem.*, 1997, **623**, 1113-1120.
63. T. y. A. Peganova, A. V. Valyaeva, A. M. Kalsin, P. V. Petrovskii, A. O. Borissova, K. A. Lyssenko and N. A. Ustynyuk, *Organometallics*, 2009, **28**, 3021-3028.
64. A. K. Gupta, F. A. S. Chipem and R. Boomishankar, *Dalton Trans.*, 2012, **41**, 1848-1853.
65. (a) M. Witt, D. Stalke, T. Henkel, H. W. Roesky and G. M. Sheldrick, *J. Chem. Soc., Dalton Trans.*, 1991, 663-667; (b) C. Qi and S. Zhang, *Appl. Organomet. Chem.*, 2006, **20**, 70-73; (c) M. Witt, H. W. Roesky, D. Stalke, F. Pauer, T. Henkel and G. M. Sheldrick, *J. Chem. Soc., Dalton Trans.*, 1989, 2173-2177; (d) E. Müller, J. Müller, F. Olbrich, W. Brüser, W. Knapp, D. Abeln and F. T. Edelmann, *Eur. J. Inorg. Chem.*, 1998, 87-91; (e) E. Rivard, C. H. Honeyman, A. R. McWilliams, A. J. Lough and I. Manners, *Inorg. Chem.*, 2001, **40**, 1489-1495.
66. K. Albahily, V. Fomitcheva, S. Gambarotta, I. Korobkov, M. Murugesu and S. I. Gorelsky, *J. Am. Chem. Soc.*, 2011, **133**, 6380-6387.
67. (a) R. Vollmerhaus, P. Shao, N. J. Taylor and S. Collins, *Organometallics*, 1999, **18**, 2731-2733; (b) R. Vollmerhaus, R. Tomaszewski, P. Shao, N. J. Taylor, K. J. Wiacek, S. P. Lewis, A. Al-Humydi and S. Collins, *Organometallics*, 2005, **24**, 494-507; (c) R. Tomaszewski, R. Vollmerhaus, A. Al-Humydi, Q. Wang, N. J. Taylor and S. Collins, *Can. J. Chem.*, 2006, **84**, 214-224.
68. A. Recknagel, M. Witt and F. T. Edelmann, *J. Organomet. Chem.*, 1989, **371**, C40-C44.
69. H. Schumann, J. Winterfeld, H. Hemling, F. E. Hahn, P. Reich, K.-W. Brzezinka, F. T. Edelmann, U. Kilimann, M. Schäfer and R. Herbst-Irmer, *Chem. Ber.*, 1995, **128**, 395-404.
70. A. Recknagel, A. Steiner, M. Noltemeyer, S. Brooker, D. Stalke and F. T. Edelmann, *J. Organomet. Chem.*, 1991, **414**, 327-335.
71. (a) B. Liu, L. Li, G. Sun, J. Liu, M. Wang, S. Li and D. Cui, *Macromolecules*, 2014, **47**, 4971-4978; (b) S. Li, D. Cui, D. Li and Z. Hou, *Organometallics*, 2009, **28**, 4814-4822.
72. (a) B. Liu, G. Sun, S. Li, D. Liu and D. Cui, *Organometallics*, 2015, **34**, 4063-4068; (b) R. Shannon, *Acta Crystallogr., Sect. A*, 1976, **32**, 751-767.
73. S. Bambirra, M. W. Bouwkamp, A. Meetsma and B. Hessen, *J. Am. Chem. Soc.*, 2004, **126**, 9182-9183.
74. K. A. Rufanov, N. K. Pruß and J. Sundermeyer, *Dalton Trans.*, 2016, **45**, 1525-1538.

75. I. A. Jaffe, K. Altman and P. Merryman, *J Clin Invest*, 1964, **43**, 1869-1873.
76. P. Benndorf, J. Jenter, L. Zielke and P. W. Roesky, *Chem. Commun.*, 2011, **47**, 2574-2576.
77. C. Averbuj, E. Tish and M. S. Eisen, *J. Am. Chem. Soc.*, 1998, **120**, 8640-8646.
78. A. Bertogg and A. Togni, *Organometallics*, 2006, **25**, 622-630.
79. J. Li, S. Huang, L. Weng and D. Liu, *J. Organomet. Chem.*, 2006, **691**, 3003-3010.
80. (a) H. Brunner and G. Agrifoglio, *Monatsh. Chem.*, 1980, **111**, 275-287; (b) H. Brunner, G. Agrifoglio, I. Bernal and M. W. Creswick, *Angew. Chem. Int. Ed.*, 1980, **19**, 641-642; (c) H. Brunner, J. Lukassek and G. Agrifoglio, *J. Organomet. Chem.*, 1980, **195**, 63-76.
81. T. Hasegawa, K. Morino, Y. Tanaka, H. Katagiri, Y. Furusho and E. Yashima, *Macromolecules*, 2006, **39**, 482-488.
82. (a) J. Kratsch, M. Kuzdrowska, M. Schmid, N. Kazeminejad, C. Kaub, P. Oña-Burgos, S. M. Guillaume and P. W. Roesky, *Organometallics*, 2013, **32**, 1230-1238; (b) M. He, Z. Chen, E. M. Pineda, X. Liu, E. Bouwman, M. Ruben and P. W. Roesky, *Eur. J. Inorg. Chem.*, 2016, 5512-5518; (c) T. P. Seifert, T. S. Brunner, T. S. Fischer, C. Barner-Kowollik and P. W. Roesky, *Organometallics*, 2018, **37**, 4481-4487; (d) P. Benndorf, J. Kratsch, L. Hartenstein, C. M. Preuss and P. W. Roesky, *Chem. Eur. J.*, 2012, **18**, 14454-14463.
83. M. He, X. Chen, T. Bodenstein, A. Nyvang, S. F. M. Schmidt, Y. Peng, E. Moreno-Pineda, M. Ruben, K. Fink, M. T. Gamer, A. K. Powell and P. W. Roesky, *Organometallics*, 2018, **37**, 3708-3717.
84. N. Li and B.-T. Guan, *Adv. Synth. Catal.*, 2017, **359**, 3526-3531.
85. K. A. Rufanov, I. Y. Titov, A. R. Petrov, K. Harms and J. Sundermeyer, *Z. Anorg. Allg. Chem.*, 2019, **645**, 559-563.
86. H. C. Brown and B. C. S. Rao, *J. Am. Chem. Soc.*, 1956, **78**, 2582-2588.
87. D. J. Parks, W. E. Piers and G. P. A. Yap, *Organometallics*, 1998, **17**, 5492-5503.
88. K. Hiromichi, I. Kazushi and N. Yoichiro, *Chem. Lett.*, 1975, **4**, 1095-1096.
89. D. Männig and H. Nöth, *Angew. Chem. Int. Ed.*, 1985, **24**, 878-879.
90. M. L. Shegavi and S. K. Bose, *Catal. Sci. Technol.*, 2019, **9**, 3307-3336.
91. (a) H. C. Brown and B. C. S. Rao, *J. Am. Chem. Soc.*, 1956, **78**, 5694-5695; (b) J. Magano and J. R. Dunetz, *Org. Process Res. Dev.*, 2012, **16**, 1156-1184; (c) J. E. Johnson, R. H. Blizzard and H. W. Carhart, *J. Am. Chem. Soc.*, 1948, **70**, 3664-3665.
92. (a) L. Fohlmeister and A. Stasch, *Chem. Eur. J.*, 2016, **22**, 10235-10246; (b) S. Yadav, S. Pahar and S. S. Sen, *Chem. Commun.*, 2017, **53**, 4562-4564; (c) J. Li, M. Luo, X. Sheng, H. Hua, W. Yao, S. A. Pullarkat, L. Xu and M. Ma, *Org. Chem. Front.*, 2018, **5**, 3538-3547; (d) S. Yadav, R. Dixit, M. K. Bisai, K. Vanka and S. S. Sen, *Organometallics*, 2018, **37**, 4576-4584; (e) M. Luo, J. Li, Q. Xiao, S. Yang, F. Su and M. Ma, *J. Organomet. Chem.*, 2018, **868**, 31-35.
93. A. P., P., de Paula, J. Atkins, *Physical Chemistry*, Oxford University Press, 2006.
94. Z. Yang, Z. Mao, Z. Xie, Y. Zhang, S. Liu, J. Zhao, J. Xu, Z. Chi and M. P. Aldred, *Chem. Soc. Rev.*, 2017, **46**, 915-1016.
95. R. Delorme and F. Perrin, *J. Phys. Radium*, 1929, **10**, 177-186.
96. G. N. Lewis, D. Lipkin and T. T. Magel, *J. Am. Chem. Soc.*, 1941, **63**, 3005-3018.
97. C. A. Parker and C. G. Hatchard, *J. Chem. Soc. Faraday Trans.*, 1961, **57**, 1894-1904.
98. H. Uoyama, K. Goushi, K. Shizu, H. Nomura and C. Adachi, *Nature*, 2012, **492**, 234-238.
99. S. A. V. Slyke, C. H. Chen and C. W. Tang, *Appl. Phys. Lett.*, 1996, **69**, 2160-2162.
100. A. Endo, M. Ogasawara, A. Takahashi, D. Yokoyama, Y. Kato and C. Adachi, *Adv. Mater.*, 2009, **21**, 4802-4806.
101. (a) R. Braveenth, H. Lee, S. Kim, K. Raagulan, S. Kim, J. H. Kwon and K. Y. Chai, *J. Mater. Chem*, 2019, **7**, 7672-7680; (b) B. M. Bell, T. P. Clark, T. S. De Vries, Y. Lai, D. S. Laitar, T. J. Gallagher, J.-H. Jeon, K. L. Kearns, T. McIntire, S. Mukhopadhyay, H.-Y. Na, T. D. Paine and A. A. Rachford, *Dyes Pigment.*, 2017, **141**, 83-92.
102. S. Bestgen, C. Schoo, B. L. Neumeier, T. J. Feuerstein, C. Zovko, R. Köppe, C. Feldmann and P. W. Roesky, *Angew. Chem. Int. Ed.*, 2018, **57**, 14265-14269.

103. (a) T. J. Feuerstein, Dissertation, Karlsruhe Institute of Technology (KIT), 2019; (b) T. J. Feuerstein, B. Goswami, P. Rauthe, R. Köppe, S. Lebedkin, M. M. Kappes and P. W. Roesky, *Chem. Sci.*, 2019, **10**, 4742-4749.
104. L. P. Spencer, R. Altwer, P. Wei, L. Gelmini, J. Gauld and D. W. Stephan, *Organometallics*, 2003, **22**, 3841-3854.
105. R. P. Kamalesh Babu, S. S. Krishnamurthy and M. Nethaji, *Tetrahedron: Asymmetry*, 1995, **6**, 427-438.
106. R. M. Ceder, C. García, A. Grabulosa, F. Karipcin, G. Muller, M. Rocamora, M. Font-Bardía and X. Solans, *J. Organomet. Chem.*, 2007, **692**, 4005-4019.
107. T. L. Troyer, H. Muchalski, K. B. Hong and J. N. Johnston, *Org. Lett.*, 2011, **13**, 1790-1792.
108. D. L. Clark, J. C. Gordon, J. C. Huffman, R. L. Vincent-Hollis, J. G. Watkin and B. D. Zwick, *Inorg. Chem.*, 1994, **33**, 5903-5911.
109. M. Rahm, R. Hoffmann and N. W. Ashcroft, *Chem. Eur. J.*, 2016, **22**, 14625-14632.
110. A. J. Wooles, M. Gregson, S. Robinson, O. J. Cooper, D. P. Mills, W. Lewis, A. J. Blake and S. T. Liddle, *Organometallics*, 2011, **30**, 5326-5337.
111. B. Wrackmeyer, C. Schödel, R. Kempe, G. Glatz and A. Noor, *Z. Anorg. Allg. Chem.*, 2016, **642**, 922-924.
112. J. C. de Mello, H. F. Wittmann and R. H. Friend, *Adv. Mater.*, 1997, **9**, 230-232.
113. (a) A. Endo, K. Sato, K. Yoshimura, T. Kai, A. Kawada, H. Miyazaki and C. Adachi, *Appl. Phys. Lett.*, 2011, **98**, 083302; (b) F. B. Dias, T. J. Penfold and A. P. Monkman, *Methods and Applications in Fluorescence*, 2017, **5**, 012001.
114. M. J. Leitl, F.-R. Küchle, H. A. Mayer, L. Wesemann and H. Yersin, *J. Phys. Chem. A*, 2013, **117**, 11823-11836.
115. Y. Cakmak, S. Kolemen, S. Duman, Y. Dede, Y. Dolen, B. Kilic, Z. Kostereli, L. T. Yildirim, A. L. Dogan, D. Guc and E. U. Akkaya, *Angew. Chem. Int. Ed.*, 2011, **50**, 11937-11941.
116. G. W. a. P. L. G. F.A.Cotton, *Basic Inorganic Chemistry*, 3rd Ed. edn., 1995.
117. F. A. Cotton, L. M. Daniels, L. R. Falvello, J. H. Matonic and C. A. Murillo, *Inorg. Chim. Acta*, 1997, **256**, 269-275.
118. R. J. Schwamm and M. P. Coles, *Organometallics*, 2013, **32**, 5277-5280.
119. (a) M. R. Crimmin, A. G. M. Barrett, M. S. Hill and P. A. Procopiou, *Org. Lett.*, 2007, **9**, 331-333; (b) M. K. Barman, A. Baishya and S. Nembenna, *J. Organomet. Chem.*, 2015, **785**, 52-60; (c) B. M. Day, N. E. Mansfield, M. P. Coles and P. B. Hitchcock, *Chem. Commun.*, 2011, **47**, 4995-4997; (d) B. M. Day, W. Knowelden and M. P. Coles, *Dalton Trans.*, 2012, **41**, 10930-10933.
120. (a) B. M. Chamberlain, M. Cheng, D. R. Moore, T. M. Ovitt, E. B. Lobkovsky and G. W. Coates, *J. Am. Chem. Soc.*, 2001, **123**, 3229-3238; (b) J. García-Álvarez, S. E. García-Garrido and V. Cadierno, *J. Organomet. Chem.*, 2014, **751**, 792-808; (c) Y. Wang, W. Zhao, X. Liu, D. Cui and E. Y. X. Chen, *Macromolecules*, 2012, **45**, 6957-6965; (d) M. H. Chisholm, J. C. Gallucci and K. Phomphrai, *Inorg. Chem.*, 2004, **43**, 6717-6725.
121. (a) M. Arrowsmith, M. S. Hill and G. Kociok-Köhn, *Chem. Eur. J.*, 2013, **19**, 2776-2783; (b) M. Arrowsmith, T. J. Hadlington, M. S. Hill and G. Kociok-Köhn, *Chem. Commun.*, 2012, **48**, 4567-4569; (c) M. Arrowsmith, M. R. Crimmin, A. G. M. Barrett, M. S. Hill, G. Kociok-Köhn and P. A. Procopiou, *Organometallics*, 2011, **30**, 1493-1506; (d) D. Mukherjee, A. Ellern and A. D. Sadow, *Chem. Sci.*, 2014, **5**, 959-964.
122. (a) M. R. Crimmin, I. J. Casely and M. S. Hill, *J. Am. Chem. Soc.*, 2005, **127**, 2042-2043; (b) M. R. Crimmin, M. Arrowsmith, A. G. M. Barrett, I. J. Casely, M. S. Hill and P. A. Procopiou, *J. Am. Chem. Soc.*, 2009, **131**, 9670-9685.
123. J. F. Dunne, S. R. Neal, J. Engelkemier, A. Ellern and A. D. Sadow, *J. Am. Chem. Soc.*, 2011, **133**, 16782-16785.
124. T. Hascall, M. M. Olmstead and P. P. Power, *Angew. Chem. Int. Ed.*, 1994, **33**, 1000-1001.
125. S. Trofimenko, *Chem. Rev.*, 1993, **93**, 943-980.

126. T. K. Panda, C. G. Hrib, P. G. Jones, J. Jenter, P. W. Roesky and M. Tamm, *Eur. J. Inorg. Chem.*, 2008, 4270-4279.
127. A. G. M. Barrett, M. R. Crimmin, M. S. Hill, P. B. Hitchcock and P. A. Procopiou, *Dalton Trans.*, 2008, 4474-4481.
128. (a) S. Qayyum, A. Noor, G. Glatz and R. Kempe, *Z. Anorg. Allg. Chem.*, 2009, **635**, 2455-2458; (b) S. Yao, H.-S. Chan, C.-K. Lam and H. K. Lee, *Inorg. Chem.*, 2009, **48**, 9936-9946.
129. (a) M. Ma, J. Li, X. Shen, Z. Yu, W. Yao and S. A. Pullarkat, *RSC Adv.*, 2017, **7**, 45401-45407; (b) D. Mukherjee, H. Osseili, T. P. Spaniol and J. Okuda, *J. Am. Chem. Soc.*, 2016, **138**, 10790-10793; (c) K. Manna, P. Ji, F. X. Greene and W. Lin, *J. Am. Chem. Soc.*, 2016, **138**, 7488-7491.
130. M. He, M. T. Gamer and P. W. Roesky, *Organometallics*, 2016, **35**, 2638-2644.
131. (a) H. Yersin, *Highly Efficient OLEDs: Materials Based on Thermally Activated Delayed Fluorescence*, John Wiley and Sons, 2019; (b) R. Czerwieniec, J. Yu and H. Yersin, *Inorg. Chem.*, 2011, **50**, 8293-8301.
132. A. Barbieri, G. Accorsi and N. Armaroli, *Chem. Commun.*, 2008, 2185-2193.
133. J. Nitsch, F. Lacemon, A. Lorbach, A. Eichhorn, F. Cisnetti and A. Steffen, *Chem. Commun.*, 2016, **52**, 2932-2935.
134. (a) L. Bergmann, C. Braun, M. Nieger and S. Bräse, *Dalton Trans.*, 2018, **47**, 608-621; (b) C.-W. Hsu, C.-C. Lin, M.-W. Chung, Y. Chi, G.-H. Lee, P.-T. Chou, C.-H. Chang and P.-Y. Chen, *J. Am. Chem. Soc.*, 2011, **133**, 12085-12099; (c) R. Czerwieniec, T. Hofbeck, O. Crespo, A. Laguna, M. Concepción Gimeno and H. Yersin, *Inorg. Chem.*, 2010, **49**, 3764-3767; (d) K. A. Barakat, T. R. Cundari and M. A. Omary, *J. Am. Chem. Soc.*, 2003, **125**, 14228-14229.
135. (a) N. Armaroli, *Chem. Soc. Rev.*, 2001, **30**, 113-124; (b) D. V. Scaltrito, D. W. Thompson, J. A. O'Callaghan and G. J. Meyer, *Coord. Chem. Rev.*, 2000, **208**, 243-266; (c) D. R. McMillin and K. M. McNett, *Chem. Rev.*, 1998, **98**, 1201-1220; (d) A. C. Lane, M. V. Vollmer, C. H. Laber, D. Y. Melgarejo, G. M. Chiarella, J. P. Fackler, X. Yang, G. A. Baker and J. R. Walensky, *Inorg. Chem.*, 2014, **53**, 11357-11366.
136. (a) H. Ohara, A. Kobayashi and M. Kato, *Dalton Trans.*, 2014, **43**, 17317-17323; (b) D. M. Zink, D. Volz, T. Baumann, M. Mydlak, H. Flügge, J. Friedrichs, M. Nieger and S. Bräse, *Chem. Mater.*, 2013, **25**, 4471-4486.
137. H.-J. Son, W.-S. Han, J.-Y. Chun, B.-K. Kang, S.-N. Kwon, J. Ko, S. J. Han, C. Lee, S. J. Kim and S. O. Kang, *Inorg. Chem.*, 2008, **47**, 5666-5676.
138. T. J. Feurestein, Dissertation, Karlsruhe Institute of Technology (KIT), 2019.
139. (a) A. Jana, T. Weyhermüller and S. Mohanta, *CrystEngComm*, 2013, **15**, 4099-4106; (b) I. Casanova, M. L. Durán, J. Viqueira, A. Sousa-Pedrares, F. Zani, J. A. Real and J. A. García-Vázquez, *Dalton Trans.*, 2018, **47**, 4325-4340.
140. B. Narayan, K. Nagura, T. Takaya, K. Iwata, A. Shinohara, H. Shinmori, H. Wang, Q. Li, X. Sun, H. Li, S. Ishihara and T. Nakanishi, *Phys. Chem. Chem. Phys.*, 2018, **20**, 2970-2975.
141. Z. Chen, A. Lohr, C. R. Saha-Möller and F. Würthner, *Chem. Soc. Rev.*, 2009, **38**, 564-584.
142. S. W. Samuel Dagorne, *Modern Organoaluminum Reagents: Preparation, Structure, Reactivity and Use*, Springer-Verlag Berlin Heidelberg, 2013.
143. S. Dagorne, I. A. Guzei, M. P. Coles and R. F. Jordan, *J. Am. Chem. Soc.*, 2000, **122**, 274-289.
144. B. Goswami, R. Yadav, C. Schoo and P. W. Roesky, *Dalton Trans.*, 2020, **49**, 675-681.
145. M.-A. Muñoz-Hernández, T. S. Keizer, S. Parkin, B. Patrick and D. A. Atwood, *Organometallics*, 2000, **19**, 4416-4421.
146. (a) R. J. Baker, C. Jones, P. C. Junk and M. Kloth, *Angew. Chem. Int. Ed.*, 2004, **43**, 3852-3855; (b) M. Zhong, Y. Liu, S. Kundu, N. Graw, J. Li, Z. Yang, R. Herbst-Irmer, D. Stalke and H. W. Roesky, *Inorg. Chem.*, 2019, **58**, 10625-10628.
147. A. L. Brazeau, Z. Wang, C. N. Rowley and S. T. Barry, *Inorg. Chem.*, 2006, **45**, 2276-2281.
148. (a) R. A. Collins, A. F. Russell and P. Mountford, *Appl. Petrochem. Res.*, 2015, **5**, 153-171; (b) G. J. P. Britovsek, V. C. Gibson and D. F. Wass, *Angew. Chem. Int. Ed.*, 1999, **38**, 428-447.

149. H. W. Roesky, B. Meller, M. Noltemeyer, H.-G. Schmidt, U. Scholz and G. M. Sheldrick, *Chem. Ber.*, 1988, **121**, 1403-1406.
150. (a) L. A. Koterwas, J. C. Fettinger and L. R. Sita, *Organometallics*, 1999, **18**, 4183-4190; (b) T. S. Brunner, L. Hartenstein and P. W. Roesky, *J. Organomet. Chem.*, 2013, **730**, 32-36.
151. S. Agarwal, C. Mast, K. Dehnicke and A. Greiner, *Macromol. Rapid Commun.*, 2000, **21**, 195-212.
152. T. G. Wetzel, S. Dehnen and P. W. Roesky, *Angew. Chem. Int. Ed.*, 1999, **38**, 1086-1088.
153. T. Gröb, G. Seybert, W. Massa, F. Weller, R. Palaniswami, A. Greiner and K. Dehnicke, *Angew. Chem. Int. Ed.*, 2000, **39**, 4373-4375.
154. M. T. Gamer, S. Dehnen and P. W. Roesky, *Organometallics*, 2001, **20**, 4230-4236.
155. U. Reißmann, P. Poremba, M. Noltemeyer, H.-G. Schmidt and F. T. Edelmann, *Inorg. Chim. Acta*, 2000, **303**, 156-162.
156. T. K. Panda, M. T. Gamer and P. W. Roesky, *Inorg. Chem.*, 2006, **45**, 910-916.
157. T. S. Brunner, P. Benndorf, M. T. Gamer, N. Knöfel, K. Gugau and P. W. Roesky, *Organometallics*, 2016, **35**, 3474-3487.
158. M. He, Z. Chen, E. M. Pineda, X. Liu, E. Bouwman, M. Ruben and P. W. Roesky, *Eur. J. Inorg. Chem.*, 2016, **2016**, 5505-5505.
159. A. J. Arduengo, R. L. Harlow and M. Kline, *J. Am. Chem. Soc.*, 1991, **113**, 361-363.
160. (a) D. Bourissou, O. Guerret, F. P. Gabbaï and G. Bertrand, *Chem. Rev.*, 2000, **100**, 39-91; (b) W. A. Herrmann, *Angew. Chem. Int. Ed.*, 2002, **41**, 1290-1309.
161. (a) A. J. Boydston, K. A. Williams and C. W. Bielawski, *J. Am. Chem. Soc.*, 2005, **127**, 12496-12497; (b) D. J. Coady, D. M. Khramov, B. C. Norris, A. G. Tennyson and C. W. Bielawski, *Angew. Chem. Int. Ed.*, 2009, **48**, 5187-5190.
162. (a) A. Kascatan-Nebioglu, M. J. Panzner, C. A. Tessier, C. L. Cannon and W. J. Youngs, *Coord. Chem. Rev.*, 2007, **251**, 884-895; (b) K. M. Hindi, M. J. Panzner, C. A. Tessier, C. L. Cannon and W. J. Youngs, *Chem. Rev.*, 2009, **109**, 3859-3884; (c) S. Ray, R. Mohan, J. K. Singh, M. K. Samantaray, M. M. Shaikh, D. Panda and P. Ghosh, *J. Am. Chem. Soc.*, 2007, **129**, 15042-15053.
163. M. C. Perry and K. Burgess, *Tetrahedron: Asymmetry*, 2003, **14**, 951-961.
164. (a) L. H. Gade and S. Bellemin-Laponnaz, *Coord. Chem. Rev.*, 2007, **251**, 718-725; (b) V. Nair, S. Vellalath and B. P. Babu, *Chem. Soc. Rev.*, 2008, **37**, 2691-2698.
165. N. Marion, S. Díez-González and S. P. Nolan, *Angew. Chem. Int. Ed.*, 2007, **46**, 2988-3000.
166. (a) N. Deak, O. Thillaye du Boullay, I.-T. Moraru, S. Mallet-Ladeira, D. Madec and G. Nemes, *Dalton Trans.*, 2019, **48**, 2399-2406; (b) J. V. Dickschat, S. Urban, T. Pape, F. Glorius and F. E. Hahn, *Dalton Trans.*, 2010, **39**, 11519-11521.
167. M. L. Bin Ismail and C.-W. So, *Chem. Commun.*, 2019, **55**, 2074-2077.
168. X. Chen, T. Simler, R. Yadav, M. T. Gamer, R. Köppe and P. W. Roesky, *Chem. Commun.*, 2019, **55**, 9315-9318.
169. (a) F. M. Mück, J. A. Baus, A. Ulmer, C. Burschka and R. Tacke, *Eur. J. Inorg. Chem.*, 2016, 1660-1670; (b) R. S. Ghadwal, H. W. Roesky, K. Pröpper, B. Dittrich, S. Klein and G. Frenking, *Angew. Chem. Int. Ed.*, 2011, **50**, 5374-5378.
170. E. Despagnet-Ayoub, H. Gornitzka, D. Bourissou and G. Bertrand, *Eur. J. Org. Chem.*, 2003, 2039-2042.
171. (a) A. V. Protchenko, J. I. Bates, L. M. A. Saleh, M. P. Blake, A. D. Schwarz, E. L. Kolychev, A. L. Thompson, C. Jones, P. Mountford and S. Aldridge, *J. Am. Chem. Soc.*, 2016, **138**, 4555-4564; (b) G. D. Frey, V. Lavallo, B. Donnadieu, W. W. Schoeller and G. Bertrand, *Science*, 2007, **316**, 439-441.
172. S. P. Green, C. Jones, P. C. Junk, K.-A. Lippert and A. Stasch, *Chem. Commun.*, 2006, 3978-3980.

173. (a) D. C. Haagenson, D. F. Moser, L. Stahl and R. J. Staples, *Inorg. Chem.*, 2002, **41**, 1245-1253; (b) D. Kost, V. Kingston, B. Gostevskii, A. Ellern, D. Stalke, B. Walfort and I. Kalikhman, *Organometallics*, 2002, **21**, 2293-2305.
174. C. Jones, S. J. Bonyhady, N. Holzmann, G. Frenking and A. Stasch, *Inorg. Chem.*, 2011, **50**, 12315-12325.
175. W. L. Armarego, *Purification of Laboratory Chemicals*, Butterworth-Heinemann 2017.
176. F. T. Edelmann, F. Pauer, M. Wedler and D. Stalke, *Inorg. Chem.*, 1992, **31**, 4143-4146.
177. S. Neander and U. Behrens, *Z. Anorg. Allg. Chem.*, 1999, **625**, 1429-1434.
178. M. Westerhausen and W. Schwarz, *Z. Anorg. Allg. Chem.*, 1992, **609**, 39-44.
179. N. Eedugurala, Z. Wang, U. Chaudhary, N. Nelson, K. Kandel, T. Kobayashi, I. I. Slowing, M. Pruski and A. D. Sadow, *ACS Catal.*, 2015, **5**, 7399-7414.
180. S. Bagherzadeh and N. P. Mankad, *Chem. Commun.*, 2016, **52**, 3844-3846.
181. G. S. Kumar, A. Harinath, R. Narvariya and T. K. Panda, *Eur. J. Inorg. Chem.*, 2020, 467-474.
182. Z. Zhu, X. Wu, X. Xu, Z. Wu, M. Xue, Y. Yao, Q. Shen and X. Bao, *J. Org. Chem.*, 2018, **83**, 10677-10683.
183. (a) G. Sheldrick, *Acta Crystallographica Section A*, 2008, **64**, 112-122; (b) G. Sheldrick, *Acta Crystallographica Section C*, 2015, **71**, 3-8.
184. O. V. Dolomanov, L. J. Bourhis, R. J. Gildea, J. A. K. Howard and H. Puschmann, *J. Appl. Cryst.*, 2009, **42**, 339-341.

8 Appendix

8.1 Directory of Abbreviations

ca.	approximately
Calcd	calculated
Dipp	2,6-$^{i}Pr_2C_6H_3$
DFT	density functional theory
Et_2O	diethylether
h	hour
Hz	Hertz
HOMO	highest occupied molecular orbital
^{i}Pr	isopropyl
IR	infrared
K	Kelvin
Ln	lanthanides
LUMO	lowest unoccupied molecular orbital
Me	methyl
mmol	millimole
NHC	N-heterocyclic carbene
NMR	nuclear magnetic resonance
^{t}Bu	tertiary butyl
R	organic group
rt	room temperature
THF	tetrahydrofuran
TM	transition metals
XRD	X-ray diffraction

(*R*)-HPEPIA *P,P*-diphenyl-*N,N'*-bis((*R*)-1-phenylethyl)phosphinimidic amine

(*R*)-HPE^{Dipp}PIA *P,P*-diphenyl-*N*-((*R*)-1-phenylethyl),*N'*-(2',6'-diisopropylphenyl)phosphin-imidic amine

(*R*)-HNEPIA *P,P*-diphenyl-*N,N'*-bis((R)-1-Naphthylethyl)phosphinimidic amine

EI	electron ionisation
MS	mass spectroscopy
Ph	phenyl
mL	millilitre
naph	naphthyl
PL	photoluminescence
PLE	photoluminescence excitation
UV	ultraviolet
ms	millisecond
μs	microsecond
UV	ultraviolet
TADF	thermally activated delayed fluorescence
τ	decay time
nm	nano meter

8.2 NMR Abbreviations

MHz	mega hertz
d	doublet
dd	doublet of doublet
J	coupling constant
m	multiplet
ppm	parts per million
q	quartet
s	singlet
t	triplet
δ	chemical shift

8.3 IR Abbreviations

br	broad
m	medium

s strong

vs very strong

w weak

vw very weak

8.4 Directory of Compounds

1 [{(*R*)-PEPIA}$_2$K$_2$]

2 [{(*R*)-PEPIA}$_2$Rb$_2$]

3 [{(*R*)-PEPIA}$_2$Cs$_2$]/ [{(*R*)-PEPIA}Cs]$_n$

4 [{(*R*)-PE^{Dipp}PIA}$_2$Li$_2$]

5 [{(*R*)-PE^{Dipp}PIA}$_2$Na$_2$]

6 [{(*R*)-NEPIA}Li(thf)$_2$]

7 [{(*R*)-HNEPIA}$_2$Na$_2$]

8 [{(*R*)-NEPIA}$_2$K$_2$]

9 [{(*R*)-PEPIA}$_2$Ca]

10 [{(*R*)-PE^{Dipp}PIA}Mg{N(SiMe$_3$)$_2$}(thf)]

11 [{(*R*)-PE^{Dipp}PIA}$_2$Mg]

12 [{(*R*)-PE^{Dipp}PIA}$_2$Ca]

13 [{(*R*)-PE^{Dipp}PIA}$_2$Yb]

14 [{(*R*)-NEPIA}$_2$Ca]

15 [{(*R*)-PE^{Dipp}PIA}$_2$Cu$_2$]

16 [{(*R*)-PE^{Dipp}PIA}$_2$Zn]

17 [{(*R*)-NEPIA}$_2$Cu$_2$]

18 [{DippPIA}$_2$Cu$_2$]

19 [{DippPIA}ZnPh]

20 [{(*R*)-PE^{Dipp}PIA}AlMe$_2$]

21 [{(*R*)-PE^{Dipp}PIA}AlMe(C$_6$F$_5$)]

22 [{(*R*)-PE^{Dipp}PIA}AlCl$_2$]

23 [{(*R*)-PEPIA}$_2$AlCl]

24 [{(*R*)-PEPIA}$_2$GaCl]

25 [{(*R*)-PEPIA}$_2$Al]$^+$[GaCl$_4$]$^-$

26	[{(*R*)-PEPIA}$_2$Ga]$^+$[AlCl$_4$]$^-$
27	[{(*R*)-PEPIA}$_2$AlH]
28	[{(*R*)-PEPIA}Hf(NMe$_2$)$_3$]
29	[{(*R*)-PE^{Dipp}PIA}Zr(NMe$_2$)$_3$]
30	[{(*R*)-PE^{Dipp}PIA}Hf(NMe$_2$)$_3$]
31	[{(*R*)-PEPIA}$_3$Y]
32	[{(*R*)-PEPIA}$_3$La]
33	[{(*R*)-PEPIA}$_3$Tb]
34	[{(*R*)-PEPIA}$_3$Yb]
35	[{(*R*)-PEPIA}$_3$Lu]
36	[{PhC(N^{t}Bu)$_2$}Si{N(*R*)(CH)MePh}{(=NDipp)PPh$_2$}]
37	[{(*R*)-PE^{Dipp}PIA}GeCl]
38	[{PhC(N^{t}Bu)$_2$}{(*S*)-PEBA}Si]
39	[{(*S*)-PEBA}GeCl]

9 Curriculum Vitae

Name	Bhupendra Goswami
Date of Birth	14th December 1993
Place of Birth	Ajmer, Rajasthan, India
Gender	Male
Marital Status	Single
Nationality	Indian

Education

Ph. D. in Chemistry	(August 2017 - present) Department of Chemistry, Karlsruhe Institute of Technology, Karlsruhe, Germany
Thesis Supervisor	Prof. Dr. Peter W. Roesky
Thesis Title	Enantiopure Iminophosphonamide Complexes: Synthesis, Photoluminescence and Catalysis
Bachelor and Master of Science	(2017) Indian Institute of Science Education and Research, Mohali, India

Poster

Alkali metal complexes of an enantiopure iminophosphonamide ligand with bright delayed fluorescence

Bhupendra Goswami, Thomas J. Feuerstein, Pascal Rauthe, Ralf Köppe, Sergei Lebedkin, Manfred M. Kappes, and Peter W. Roesky

The 13th International Conference on Heteroatom Chemistry (**ICHAC 2019**) in Prague from June 30 to July 5, 2019

Publications

1) D. Bawari, B. Goswami, S.K. Thakur, R.V.V. Tej, A.R. Choudhury, S. Singh. *Dalton Trans.*, **2018**, *47*, 6274-6278.

2) T. J. Feuerstein, B. Goswami, P. Rauthe, R. Köppe, S. Lebedkin, M. M. Kappes, P.W. Roesky. *Chem. Sci.*, **2018**, *10*, 4742-4749.

3) D. Bawari, S. K. Thakur, K. K. Manar, B. Goswami, Sabari V. R., A. R. Choudhury, S. Singh

J. Organomet. Chem. **2019,** *880,* 108-115.

4) B. Goswami, R. Yadav, C. Schoo, P.W. Roesky, *Dalton Trans.*, **2020**, *49*, 675-681.

5) R. Yadav, T. Simler, S. Reichl, B. Goswami, Dr C. Schoo, R. Köppe, M. Scheer, P.W. Roesky, *J. Am. Chem. Soc.* **2020**, *142*, 1190-1195.

6) R. Yadav, T. Simler, B. Goswami, C. Schoo, R. Köppe, S. Dey, and P. W. Roesky, *Angew. Chem. Int. Ed.* **2020**, *59*, 9443-9447.

10 Acknowledgements

First of all, I would like to express my sincere gratitude towards my supervisor Prof. Dr. Peter W. Roesky for his generous support, constant encouragement, and excellent guidance during every stage of my doctoral thesis. His unwavering enthusiasm on the different areas of inorganic chemistry and very honest and constructive criticism of my work helped me to understand various aspects of synthetic organometallic chemistry. In addition, his friendly nature and easy-going attitude made my stay pleasant here. I am very grateful for getting an opportunity to work in his research group.

I would like to thank Dr. Ralf Köppe for his help with theoretical calculations.

I am very thankful to my very good friend Dr. Ravi Yadav for his constant encouragement and help throughout the PhD, it has been a great experience working with him. I am thankful to my good friend Xiaofei Sun for being amazing colleague and providing a wonderful working environment.

I am very thankful to Sibylle Schneider for measuring crystals and for her very kind and helpful nature. I would like to thank Dr. Michael Gamer, Dr. Thomas Simler and Dr. Christoph Schoo for their help in solving the crystal structures.

I appreciate Mrs. Kayas and Mrs. Pendl for their assistance with paper and document work. I am also especially thankful to Mrs. Berberich (NMR), Mrs. Smie (Mass), Mrs. Stößer (chemical ordering), Mrs. Klaassen (elemental analysis), Mrs. Leichle (chemical store), Mr. Munshi (glassblowing), Mr. Rieß, Mr. Lampert, and Mr. Kastner (mechanical workshop) for their help.

I would like to thank Dr. Ravi Yadav, Dr. Tim Seifert, Dr. Thomas Feuerstein, Xiaofei Sun, and Dr. Thomas Simler for proofreading my thesis. I would like to thank Niklas Reinfandt, Dr. Thomas Feuerstein and Dr. Sebastian Kaufmann for proofreading and German translation of my summary. I am thankful to Dr. Sebastian Kaufmann for helping me in software related problems. I would like to thank all of my colleagues for their help and suggestions during my work.

I would like to thank my friends Omprakash, Ravi, Vikram, Navid, Tarachand, Krishan for their constant encouragement and support.

Finally, I would like to thank my parents, my brother, my sisters and my whole family for believing in me and for their continuous support and encouragement which made my work possible.

www.ingramcontent.com/pod-product-compliance
Ingram Content Group UK Ltd.
Pitfield, Milton Keynes, MK11 3LW, UK
UKHW022001190726
13853UKWH00004B/1665

9 783736 973275